Euclid's "Elements" Redux
Vol. 1: Plane Geometry

Daniel Callahan, John Casey, Sir Thomas Heath

April 23, 2022

"Euclid's 'Elements' Redux" ©2022 Daniel Callahan, licensed under the Creative Commons Attribution-ShareAlike 4.0 International License.

Selections from the American Heritage® Dictionary of the English Language, Fourth Edition, ©2000 by Houghton Mifflin Company. Updated in 2009. Published by Houghton Mifflin Company. All rights reserved.

Selections from Dictionary.com Unabridged Based on the Random House Dictionary, © Random House, Inc. 2013. All rights reserved.

Wikipedia® is a registered trademark of the Wikimedia Foundation, Inc., a non-profit organization.

Questions? Comments? Did you find an error?

Email me at: dpcallahan@protonmail.com

Make sure to include the version number: 2022-113

Download this book for free:

https://archive.org/details/euclid-elements-redux_201809

Also by Daniel Callahan:

215 Solutions to Problems from Linear Algebra
5th ed., Friedberg, Insel, Spence

266 Solutions to Problems from Linear Algebra
4th ed., Friedberg, Insel, Spence

Amazing Roleplaying Games' 193Q

Any Coincidence Is

Euclid's Elements Redux Vol. 1: Plane Geometry
(Books I-VI, print)

Euclid's Elements Redux Vol. 2: Number Theory
(Books VII-X, print)

Fēlēs et Canis (Latin)
Feles et Duo Canes (Latin)
The Purr-fect Cat and her Stupid Dogs (Latin-English Bilingual Edition)

"Don't just read it; fight it! Ask your own questions, look for your own examples, discover your own proofs. Is the hypothesis necessary? Is the converse true? What happens in the classical special case? What about the degenerate cases? Where does the proof use the hypothesis?"
- Paul Halmos

"Pure mathematics is, in its way, the poetry of logical ideas."
- Albert Einstein

"Math is like going to the gym for your brain. It sharpens your mind."
- Danica McKellar

"One of the endlessly alluring aspects of mathematics is that its thorniest paradoxes have a way of blooming into beautiful theories."
- Philip J. Davis

Mathematics is a discipline, which means that you compete against yourself. If you continue to improve, then you're a mathematician.

Contents

0 About this project   **7**
     0.1   Contributions & Acknowledgments 10
     0.2   Dedication 11
     0.3   Who needs a new edition of Euclid's Elements? 11
     0.4   Why rewrite Euclid's Elements? 11
     0.5   Recommended Reading 13

1 Angles, Parallel Lines, Parallelograms   **15**
     1.1   Symbols, Logic, and Definitions 15
     1.2   Postulates 27
     1.3   Axioms 28
     1.4   Book I, Propositions 1-26 33
     1.5   Book I, Propositions 27-48 75

2 Rectangles   **117**
     2.1   Definitions 117
     2.2   Axioms 119
     2.3   Propositions from Book II 119

3 Circles   **139**
     3.1   Definitions 139
     3.2   Propositions from Book III 144

4 Inscription and Circumscription   **211**
     4.1   Definitions 211
     4.2   Propositions from Book IV 214

5 Theory of Proportions — **241**

   5.1 Definitions 241

   5.2 Propositions from Book V 243

6 Applications of Proportions — **249**

   6.1 Definitions 249

   6.2 Propositions from Book VI 252

14 Solutions — **321**

   14.1 Solutions for Chapter 1 321

   14.2 Solutions for Chapter 2 361

   14.3 Solutions for Chapter 3 365

   14.4 Solutions for Chapter 4 372

   14.5 Solutions for Chapter 5 376

   14.6 Solutions for Chapter 6 377

Chapter 0

About this project

The goal of this textbook is to provide an edition of Euclid's *Elements* that is easy to read, inexpensive, and has been released under an open culture license.[1]

The title *Euclid's "Elements" Redux* indicates that while this edition states and proves Euclid's results, the proofs have been rewritten using modern mathematics. This is because mathematics has almost completely changed since *The Elements* was published c. 300 BC.

In Euclid's Hellenistic culture, a number was synonymous with a length that could be measured. This gave Euclid access to all of the positive numbers but neither zero nor the negatives. But since the development of algebra wouldn't begin for another millennium, Euclid could not have known that the result of Book I, Proposition 47 would be summarized as $a^2 + b^2 = c^2$.

When a modern student encounters the original *Elements* for the first time, he or she encounters two problems: the logic of the proof, and Euclid's now archaic concept of numbers. This edition updates Euclid's proofs while retaining his fundamental results.

It may be impossible to overstate how fundamental these results are to mathematics. They are the primary reason why *The Elements* was the world's most important mathematics textbook for about 2,200 years.

Let that number sink in for a moment... one math textbook was used by much of the literate world for over two thousand years. Why should this be? Lack of competition? At certain times and places, yes, but in schools where Euclid's *Elements* got a foothold, other textbooks soon followed. Therefore, this can't be the complete answer.

---

[1] *Euclid's "Elements" Redux* is released under the Creative Commons Attribution-ShareAlike 4.0 International License. This means that you are free to copy and distribute it without penalty. If you wish to add to its content, your work must also be licensed under an equivalent open license.

The book and its source files are available online at
https://archive.org/details/euclid-elements-redux_201809

The figures were created in GeoGebra and can be found in the relevant images folder. Files with extensions .ggb are GeoGebra files, and files with the .eps extension are graphics files.

Elegant proofs? While Euclid's originals remain a model for how proofs should be written, they aren't irreplaceable.

Content? Until the development of algebra and calculus, *The Elements* covered everything a novice mathematician needed to know. And even after the advent of calculus, *The Elements* still had its place. Isaac Newton's assistant claimed that in five years of service, he had heard Newton laugh just once: a student asked if Euclid was still relevant, and Newton laughed at him. But most students only studied the first six Books (i.e., the first six of the thirteen scrolls that made up the original *Elements*). Therefore, not everyone studied *The Elements* exclusively for its content.

So, whatever is valuable about *The Elements* must be present in its first half; but each Book begins with a foundation (definitions and axioms) and builds upwards one proposition at a time.

This, I think, is the key to the question: not only does *The Elements* help a student learn geometry, but it also immerses the student in a logical system that is as useful as it is penetrating. While learning algebra helps a student to perform calculus and statistics, learning geometric proofs from *The Elements* helps a student to think clearly about politics, art, music, design, coding, law... any subject that requires rational thought can be better understood after Euclid.

The reason *The Elements* was needed in the past, and why it's needed today, is that it helps its readers learn to think clearly.

But if *The Elements* is a good textbook, why was it abandoned at the end of the 19th century?

Mathematics faced a crisis in the last half of that century – ambiguous definitions and sloppy logic had led to serious contradictions. Without a complete overhaul of common definitions and the construction of formal logic, mathematics would have collapsed.

After several decades of work by many brilliant minds, the overhaul was completed, leading to a mathematical golden age which is still unfolding. But that overhaul made Euclid's weaknesses clear: while his logic remained sound, the presentation of that logic was outdated. Euclid's definitions and assumptions were ambiguous to the point of being unworkable. Most seriously, other geometries had been developed, proving that Euclid's work was not unique.

As calculus developed into the foundation of engineering and the sciences, its prerequisite, algebra, became the course every student seemed to require. Geometric proofs became a luxury rather than a necessity. With perfect hindsight, the error is clear – forcing each and every student to learn algebra has only succeeded in teaching students to hate "math" (which millions of people associate exclusively with algebra) while hindering their attempts to think logically.

The way forward is also clear – either to rewrite *The Elements* or to develop a new, equivalent work. I have opted to rewrite Euclid in the hope that *The Elements* will again be recognized as a textbook of introductory, axiomatic geometry, a model of proof-writing, and a case-study in applied logic.[2]

Like all math textbooks, *Euclid's "Elements" Redux* requires its student to work slowly and carefully through each section. The student should confirm each result and not take anything on faith. While this process may seem tedious, it is exactly this attention to detail which separates those who understand mathematics from those who do not.

This edition also contains homework problems and a partial answer key. However, no prerequisites are required if the student's goal is to read and understand the material. The best way to demonstrate this understanding is to memorize a certain number of proofs and then recite them on request. This was how Abraham Lincoln made use of *The Elements*:

> In the course of my law-reading, I constantly came upon the word demonstrate. I thought at first that I understood its meaning, but soon became satisfied that I did not. I said to myself, "What do I mean when I demonstrate more than when I reason or prove? How does demonstration differ from any other proof?" ... I consulted all the dictionaries and books of reference I could find, but with no better results. You might as well have defined blue to a blind man. At last I said, "Lincoln, you can never make a lawyer if you do not understand what demonstrate means;" and I left my situation in Springfield, went home to my father's house, and stayed there till I could give any proposition in the six books of Euclid at sight. I then found out what "demonstrate" means, and went back to my law-studies.[3]

If the student intends to prove some of the problems, then proportions, algebra, trigonometry, and possibly linear algebra will be helpful. Students with no knowledge of proof writing should consult Richard *Hammack*'s "Book of Proof", 3rd ed.[4]

It is vitally important to understand that, in the mind of mathematicians, math is a collection of statements about relationships between quantities which can be proven. Without proofs, mathematics does not exist.

This document was composed over the years using a number of tools:

---

[2] Victor Aguilar has opted to develop a new introduction to axiomatic geometry. See section [0.4] for details.
[3] Carpenter, F.B. "The Inner Life of Abraham Lincoln: Six Months at the White House" Hurd & Houghton, New York, NY (1874).
[4] https://www.people.vcu.edu/~rhammack/BookOfProof/

Debian	`http://www.debian.org/`
GeoGebra	`http://www.geogebra.org/`
Kubuntu	`https://kubuntu.org`
Linux Mint	`http://www.linuxmint.com/`
LyX	`http://www.lyx.org/`
Windows 7	`http://windows.microsoft.com/`
Xubuntu	`https://xubuntu.org`

0.1 Contributions & Acknowledgments

Contributors and their contributions:

- Victor Aguilar (an invaluable second opinion)

- Daniel Callahan (general editor)

- Deirdre Callahan (corrections)

- John Casey (Casey's edition of "The Elements" is a partial basis for chapters 1-6)

- Daniel Ezell (Owner/Teacher at Golden Gate Learning Center)

- Jared Gans (corrections)

- Ralph Giles (corrections)

- Holly Haynes (contributor)

- Sir Thomas L. Heath (Heath's edition of "The Elements" is this original basis for chapter 7 and beyond)

- Domagoj Hranjec (contributor)

- Robert Jullien (corrections)

- Elizabeth B. Morran (proofreading)

- S. P. (proofreading)

- Andreas Piotrowski (moral support)

- Neven Sajko (corrections)

- Moustafa Shahin (corrections)

- Valorie Starr (proofreading)

- Contributors to proofwiki.org

Daniel Callahan would like to thank Jon Allen, Dr. Wally Axmann[5], Dr. Elizabeth Behrman, Karl Elder[6], David E. Joyce[7], Dr. Thalia Jeffres[8], Dr. Kirk Lancaster[9], Dr. Phil Parker[10], Dr. Gregory B. Sadler[11], and Valorie Starr.

0.2 Dedication

This book is dedicated to everyone in the educational community who believes that algebra provides a better introduction to mathematics than geometry.

0.3 Who needs a new edition of Euclid's Elements?

Consider an analogous question: who needs training wheels on a bike?

A young person who doesn't know how to ride a bicycle.

Experienced mathematicians may no longer need Euclid, but Euclid's construction of complex ideas from simple axioms remains a model for how mathematics should be approached. Students who attempt to master Euclid's Elements will find 21st century mathematics less confusing despite Euclid's less-than-rigorous definitions.

0.4 Why rewrite Euclid's Elements?

The concept that Euclid could use a little tweaking goes back a long way. Book I, Proposition 40 has been identified as an interpolation, along with many of *The Elements*' lemmas and corollaries. Some editions include the apocryphal Books XIV and XV which add results on the topic of solid geometry.

It's important to realize that no math textbook is perfect; flaws will inevitably come to light after centuries of close study by intelligent minds.

But it's also important to realize that more than one correct geometry exists; Euclid's geometry is one of many (but perhaps the easiest to learn). Similarly, for any true and logical result, either more than one proof exists or the potential for more than one proof exists. Euclid's proofs need not be treated as special because they are "the originals".[12]

---

[5] http://www.math.wichita.edu/~axmann/
[6] http://karlelder.com
[7] https://mathcs.clarku.edu/~djoyce/java/elements/elements.html
[8] http://www.math.wichita.edu/~jeffres/
[9] http://www.math.wichita.edu/people/lancaster.html
[10] http://www.math.wichita.edu/~pparker/
[11] http://gregorybsadler.com/
[12] Some of the proofs in this text are not based on Euclid's originals. The originals can be found in several printed editions, online at David E. Joyce's Euclid's *Elements*:
http://aleph0.clarku.edu/~djoyce/java/elements/elements.html
and in Richard Fitzpatrick's "Euclid's Elements of Geometry":
https://farside.ph.utexas.edu/Books/Euclid/Elements.pdf

There at least two reasons for this. First, a student of mathematics should always ask if there is another way to prove an interesting theorem. Doing so may provide insight, if not help generate a new result.

Second, it's doubtful that Euclid (if indeed he was a single individual) is the sole author of these proofs. It's more likely that he (or the scholars of his school) compiled and rewrote these proofs from difference sources. Rewriting and editing is part of a mathematician's work.

This is not to denigrate the achievement of "The Elements" – the original thirteen books may have been the first to demonstrate how to construct hundreds of complex structures beginning with first principles. Nearly all well-written math, physics, and engineering textbooks follow a similar format (and all of the bad ones do not).

But is still the case that Euclid's original proofs are obsolete – they refer to a conception of mathematics that is no longer viable because it cannot be extended to produce real analysis, complex analysis, etc. To help see this, consider Euclid's original proof of [1.3]:

Book I, Proposition 3: For two given unequal straight-lines, to cut off the greater a straight-line equal to the lesser.

Proof. Let AB and C be the two given unequal straight-lines, of which let the greater be AB. So it is required to cut off a straight-line equal to the lesser C from the greater AB.

Let the line AD, equal to the straight-line C, have been placed at point A. And let the circle have been drawn with center A and radius AD.

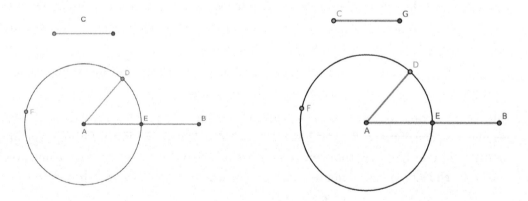

Figure 0.4.1: Book I, Proposition 3 (original on the left, rewrite on the right)

And since point A is the center of circle DEF, AE is equal to AD. But, C is also equal to AD. Thus, AE and C are each equal to AD. So AE is also equal to C.

Thus, for two given unequal straight-lines, AB and C, the (straight-line) AE, equal to the lesser C, has been cut off from the greater AB. (Which is) the very thing it was required to do.[13]

Compare the original to the rewritten proof below:

Given two arbitrary segments which are unequal in length, it is possible to subdivide the larger segment such that one of its two sub-segments is equal in length to the smaller segment.

Proof. Construct segments $\overline{AB}$ and $\overline{CG}$ such that $\overline{CG} < \overline{AB}$. We claim that $\overline{AB}$ may be subdivided into segments $\overline{AE}$ and $\overline{EB}$ where $\overline{AE} = \overline{CG}$.

From point A, construct the segment $\overline{AD}$ such that $\overline{AD} = \overline{CG}$ [1.2]. With A as the center and $\overline{AD}$ as radius, construct the circle $\odot A$ [Postulate 1.3] which intersects $\overline{AB}$ at E.

Because A is the center of $\odot A$, $\overline{AE} = \overline{AD}$ [Def. 1.32]. Since $\overline{AD} = \overline{CG}$ by construction, by Axiom 9 from section 1.3.1 (using equalities), we find that $\overline{AE} = \overline{CG}$, which proves our claim. □

0.5 Recommended Reading

Book of Proof, 3rd edition, Richard Hammack.

This open textbook provides an introduction to the standard methods of proving mathematical theorems. It can be considered a companion volume to any edition of Euclid, especially for those who are learning how to read and write mathematical proofs for the first time. It has been approved by the American Institute of Mathematics' Open Textbook Initiative and has a number of good reviews at the Mathematical Association of America Math DL and on Amazon.

`http://www.people.vcu.edu/~rhammack/BookOfProof/index.html`

Euclid's *Elements* Online, coded and maintained by David E. Joyce

`http://aleph0.clarku.edu/~djoyce/java/elements/elements.html`

Geometry Without Multiplication, Victor Aguilar

This geometry textbook will be the first in a series (the second being Geometry With Multiplication). It will not only be suitable for high school students but will also maintain the rigor required in college level textbooks. The most recent draft can be found at:

---

[13] From Richard Fitzpatrick's "Euclid's Elements of Geometry".

```
https://www.researchgate.net/publication/291333791_Volume_One_Geometry_without_
Multiplication
```

Guidelines for Good Mathematical Writing, Francis Edward Su

```
https://www.math.hmc.edu/~su/math131/good-math-writing.pdf
```

How to Solve it, George Polya

"A perennial bestseller by eminent mathematician G. Polya, *How to Solve It* will show anyone in any field how to think straight. In lucid and appealing prose, Polya reveals how the mathematical method of demonstrating a proof or finding an unknown can be of help in attacking any problem that can be 'reasoned' out – from building a bridge to winning a game of anagrams. Generations of readers have relished Polya's deft – indeed, brilliant – instructions on stripping away irrelevancies and going straight to the heart of the problem."[14]

```
http://www.amazon.com/How-Solve-Mathematical-Princeton-Science/dp/069116407X/
```

Khan Academy   `https://www.khanacademy.org/`

The King of Infinite Space: Euclid and His Elements, David Berlinski

Not an edition of Euclid's Elements but an explication of the *Elements* itself and what makes the work revolutionary.

```
https://www.amazon.com/King-Infinite-Space-Euclid-Elements-ebook/dp/B00HTQ320S
```

Math Open Reference

```
http://www.mathopenref.com/
```

```
http://www.mathopenref.com/trianglecenters.html
```

---

[14]This description (c)1985 by Princeton University Press.

Chapter 1

Angles, Parallel Lines, Parallelograms

Students should construct figures and/or work through the proofs step-by-step. This is an essential component to the learning process that cannot be avoided. The old saying, "There is no royal road to geometry", means: "No one learns math for free."

1.1 Symbols, Logic, and Definitions

The propositions of Euclid will be referred to in brackets; for example, we[1] write [3.32] instead of writing Proposition 3.32. Axioms, Definitions, etc., will also be referred to in this way; for example, Definition 12 in chapter 1 will be written as [Def. 1.12]. Exercises to problems will be written as [3.5, #1] instead of exercise 1 of Proposition 3.5.

Numbered equations will be written as (10.2.2) instead of the second equation in chapter 10, section 2.

A note on the exercises: *do some of them but don't feel pressured to do all of them.* Generally, an exercise is expected to be solved using the propositions, corollaries, and exercises that preceded it. For example, exercise [1.32, #3] should first be attempted using propositions [1.1]-[1.32] as well as all previous exercises. Should this prove too difficult or too frustrating, he or she should consider whether propositions [1.33] or later (and their exercises) might help solve the exercise. It is also permissible to use trigonometry, linear algebra, or other contemporary mathematical techniques on challenging problems.[2]

---

[1] Mathematicians often write "we" when writing about math in the same way that coaches tell players "here's what we're going to do" – we are engaged in a team effort to overcome our difficulties.

[2] And remember that math makes everyone feel stupid at times. Never give up.

CHAPTER 1. ANGLES, PARALLEL LINES, PARALLELOGRAMS

1.1.1 Symbols

The following symbols will be used to denote standard geometric shapes or relationships:

- Circles are denoted by: $\bigcirc$. When the center of a circle is known (for example, point A), the circle will be identified as $\bigcirc A$. Otherwise, the circle will be identified with points on its circumference, such as $\bigcirc ABC$.

- Triangles by: $\triangle$

- Parallelograms by: $\square$

- Parallel lines by: $\parallel$

- Perpendicular lines by: $\perp$

In addition to these, we shall employ the usual symbols of algebra: $+, -, =, <, >, \leq, \geq, \neq$, as well as a few additional symbols:

- Composition: $\oplus$ For example, suppose we have the segments $\overline{AB}$ and $\overline{BC}$ which intersect at the point B. The statement $\overline{AB} + \overline{BC}$ refers to the sum of their lengths, but $\overline{AB} \oplus \overline{BC}$ refers to their composition as one object. See Fig. 1.1.1.

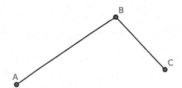

Figure 1.1.1: Composition: the geometrical object $\overline{AB} \oplus \overline{BC}$ is a single object composed of two segments, $\overline{AB}$ and $\overline{BC}$.

The composition of angles, however, can be written using either $+$ or $\oplus$, and in this textbook their composition will be written with $+$.

- Similar: $\sim$ Two figures or objects are similar if they have the same shape but not necessarily the same size. If two similar objects have the same size, they are also congruent.

- Congruence: $\cong$ Two figures or objects are congruent if they have the same shape and size, or if one has the same shape and size as the mirror image of the other. This means that an object can be re-positioned and reflected so as to coincide precisely with the other object without resizing.[3]

---

[3] http://en.wikipedia.org/wiki/Congruence_(geometry)

1.1.2 Logic

Propositions are mathematical statements that are either completely true or completely false, but never both.[4] Some examples and counterexamples:

- "A triangle has two sides" is a false proposition.
- "A triangle has three sides" is a true proposition.
- "A triangle has three sides?" is not a proposition; it is a question, not a statement.
- "Draw a triangle" is not a proposition; it is a command.

Propositions which are true may be divided into axioms and theorems. An axiom needs no proof, and a theorem requires at least one proof; both axioms and theorems are considered *true*. A proposition which cannot be proven true is neither an axiom nor a theorem and is considered *false*. (Minor spoiler: all propositions in this book will be shown to be true, i.e., they are theorems.)

An *axiom* is a proposition that is assumed to be true without proof[5]. They are considered so fundamental that they cannot be inferred from any proposition which is more elementary. "Any two sides of a triangle are greater in length than the third side" may be self-evident; however, it is not an axiom since it can be inferred by demonstration from other propositions. The statement "two objects which are equal in length to a third object are also equal in length to each other" is self-evident, and so it is considered an axiom.[6]

A *theorem* is a proposition that may be proven from known propositions (either theorems or axioms). Theorems may also be described as formal statements of mathematical or logical properties.

A *proof* is a rigorous mathematical argument which unequivocally demonstrates the truth of a given proposition[7]. A proof consists of three parts: the *hypothesis*, that which is assumed, the *claim*, that which the author intends to prove, and the bulk of the proof which demonstrates how the claim must be true once the hypothesis is assumed to be true.

A *corollary* is an inference or deduction based on a theorem which usually states a small but important result that follows immediately from the proof itself or from the

---

[4] Propositions are also never partly true. If your dog has brown fur except for one white paw, the proposition "Your dog is brown" is false.

[5] Source: Weisstein, Eric W. "Axiom." From MathWorld Wolfram Web Resource. http://mathworld.wolfram.com/Axiom.html

[6] Whether a given statement is considered an axiom or a theorem depends on which textbook you are reading. Graduate textbook authors may require students to prove statements which were considered axiomatic at the undergraduate level.

[7] Source: Weisstein, Eric W. "Proof." From MathWorld - A Wolfram Web Resource. http://mathworld.wolfram.com/Proof.html

result of the theorem. For example, if a theorem states that all prime numbers have irrational square roots, then one corollary to this theorem is that $\sqrt{2}$ is irrational.[8]

A *lemma* is a theorem which is used as a stepping stone to a larger result rather than as a statement of interest by itself.[9] While technically all lemmas are theorems, lemmas are not called theorems in order to communicate the idea that the result is of minor importance and exists to help prove something more profound.[10]

1.1.2.1 Examples

Proposition. *(1) If x is a rational number, then x has a decimal expansion.*

The hypothesis is that "x is a rational number", and the claim is that "x has a decimal expansion" (i.e., it can be written in decimal form). In order to prove this proposition is a theorem, we begin by assuming that x is a rational number (e.g., a fraction). From this, we must show logically that x has a decimal expansion. If we can do this, we have written a proof, turning this proposition into a theorem.[11]

Converse statements: if we rewrite the above proposition by swapping the hypothesis and the claim, we obtain its converse statement:

Proposition. *(2) If x has a decimal expansion, then x is a rational number.*

Since this proposition is false, it has no proof and therefore is not a theorem.

> There is no guarantee that the converse of a given proposition will be true.

From propositions (1) and (2), we may infer two others: their *contrapositive propositions* (informally called contrapositives). The contrapositive forms of proposition (1):

Proposition. *(3) If x does not have a decimal expansion, then x is not a rational number.*

The contrapositive forms of proposition (2):

Proposition. *(4) If x is not a rational number, then x does not have a decimal expansion.*

Unlike converse propositions, a contrapositive proposition is true if and only if the original proposition is true. Since (1) is true, (3) is true; since (3) is true, (1) is true. Similarly, a contrapositive proposition is false if and only if the original proposition is false: since (2) is false, (4) is false; since (4) is false, (2) is false.

---

[8]This important result will not be proven here. Interested readers should consult Dummit & Foote's "Abstract Algebra", 3rd edition.
[9]Source: https://en.wikipedia.org/wiki/Lemma_(mathematics)
[10]The difference between a lemma and a theorem is often determined by the author. For example, it could be considered that propositions 1-46 in Chapter 1 are lemmas leading toward propositions 47-48, which are the most profound statements in the chapter.
[11]While this is indeed a theorem, we will not prove it here.

1.1.3 Definitions

We need a common language in order to discuss similar experiences or ideas. For mathematics, this is true to an almost ridiculous degree. A mathematician's work can be rendered useless if the definitions he or she employs turn out to be vague or sloppy.[12]

Students reading this section for the first time may wish to read definitions 1-6 and 9-11 and then skip ahead to [1.4], returning to the remaining definitions as well as [1.2] and [1.3] as needed.

The Point

1. A *point* is a zero dimensional object.[13] A geometrical object which has three dimensions (length, height, and width) is a solid. A geometrical object which has two dimensions (length and height) is a surface, and a geometrical object which has one dimension is a line or line segment. Since a point has none of these, it has zero dimensions.

The Line

2. A *line* is a one dimensional object: it has only length. If it had any height or width, no matter how small, it would have two dimensions. Hence[14], a line has neither height nor width.

A line with points A and B is written as $\overleftrightarrow{AB}$.

(This definition conforms to Euclid's original definition in which a line need not be straight. However, in all modern geometry texts, it is understood that a "line" has no curves. See also [Def 1.4].)

3. The intersections of lines are points. However, a point may exist without being the intersection of lines.

4. A line without a curve is called a *straight line*. It is understood throughout this textbook that a *line* refers exclusively to a *straight line*. A curved line (such as the circumference of a circle) will never be referred to merely as a line in order to avoid confusion. Lines have no endpoints since they are infinite in length.

---

[12]Euclid's original definitions are almost useless to modern mathematicians. One example: Euclid defines a line as either straight or curved and either finite or infinite in length. This type of vagueness, common until the early 19th century, was one reason why mathematics had to be rewritten into its modern form.

[13]Warren Buck, Chi Woo, Giangiacomo Gerla, J. Pahikkala. "point" (version 13). PlanetMath.org. Freely available at http://planetmath.org/point

[14]"Hence", along with "thus" and "therefore", are three words that mathematicians use to mean "consequently" or "for this reason". Generally, "hence" is used for minor results, "thus" for major results, and "therefore" for results in between. However, YMMV.

A *line segment* (or more simply a *segment*) is similar to a line except that it is finite in length and has two *endpoints* at its extremities. A line segment with endpoints A and B is written as $\overline{AB}$. (If the length of a segment appears in a fraction, we will omit the overline. For example, $1 = \frac{AB}{AB}$ for any segment $\overline{AB}$.)

A *ray* is like a line in that it is infinite in length; however, it has one endpoint. A ray with endpoint A and point B is written as $\overrightarrow{AB}$ (where A is the endpoint).

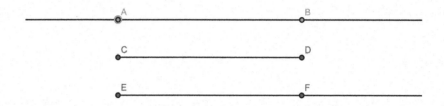

Figure 1.1.2: [Def. 1.2, 1.3, 1.4] $\overleftrightarrow{AB}$ is a line, $\overline{CD}$ is a segment (sometimes called a line segment), and $\overrightarrow{EF}$ is a ray

The Plane

5. A *surface* has two dimensions: length and height. It has no width; if it had, it would be a space of three dimensions.

6. A *plane* is a surface that extends infinitely far and is assumed to be completely flat. A plane is the two-dimensional analogue of a point (zero dimensions), a line (one dimension) and three-dimensional space. Planes act as the setting for most of Euclidean geometry; that is, "the plane" refers to the whole space in which two-dimensional geometry is performed.

Planes are defined by three points. For any three points not on the same line, there exists one and only one plane which contains all three points.

7. Any combination of points, lines, line segments, or curves on a plane is called a *plane figure*. A plane figure that is bounded by a finite number of straight line segments closed in a loop to form a closed chain or circuit is called a *polygon*[15].

All bounded plane figures have a measure called *area*. Area[16] is the quantity that expresses the extent of a two-dimensional figure or shape on a plane. Area can be understood as the amount of material with a given thickness that would be necessary to fashion a model of the shape, or the amount of paint necessary to cover the surface with a single coat.

---

[15] http://en.wikipedia.org/wiki/Polygon
[16] Taken from the article: https://en.wikipedia.org/wiki/Area.

1.1. SYMBOLS, LOGIC, AND DEFINITIONS

Area is the two-dimensional analog of the length of a curve (a one-dimensional concept) or the volume of a solid (a three-dimensional concept). Surface area is its analog on the two-dimensional surface of a three-dimensional object.[17]

8. Points which lie on the same straight line, ray, or segment are called *collinear points*.

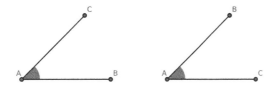

Figure 1.1.3: [Def 1.11] Notice that both angles could be referred to as $\angle BAC$, $\angle CAB$, or the angle at point A where A is a vertex.

The Angle

9. The angle made by two straight lines, segments, or rays extending outward from a common point but in different directions is called a *rectilinear angle* (or simply an *angle*).

10. The one point of intersection between straight lines, rays, or segments is called the *vertex of the angle*.

11. A particular angle in a figure will be written as the symbol $\angle$ and three letters, such as BAC, of which the middle letter, A, is at the vertex. Hence, such an angle may be referred to either as $\angle BAC$ or $\angle CAB$. Occasionally, this notation will be shortened to "the angle at point A" instead of naming the angle as above.

12. The angle formed by composing two or more angles is called their sum. Thus in Fig. 1.1.4, we find that $\angle ABC \oplus \angle PQR = \angle ABR$ where the segment $\overline{QP}$ is applied to the segment $\overline{BC}$. We generally write $\angle ABC + \angle PQR = \angle ABR$ to express this concept.

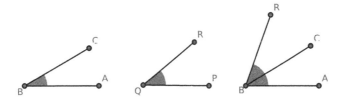

Figure 1.1.4: [Def. 1.12]

---

[17] See the List of Formulas in: https://en.wikipedia.org/wiki/Area

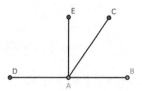

Figure 1.1.5: [Def. 1.13]

13. Suppose two segments $\overline{BA}$, $\overline{AD}$ are composed such that $\overline{BA} \oplus \overline{AD} = \overline{BD}$ where $\overline{BD}$ is a segment (see Fig. 1.1.5). If a point C which is not on the segment $\overline{BD}$ is connected to point A, then the angles $\angle BAC$ and $\angle CAD$ are called *supplements* of each other. This definition holds when we replace segments by straight lines or rays, *mutatis mutandis*[18].

14. When one segment, $\overline{AE}$, stands on another segment, $\overline{BD}$, such that the adjacent angles on either side of $\overline{AE}$ are equal (that is, $\angle EAD = \angle EAB$), each of the angles is called a *right angle*, and the segment which stands on the other is described as *perpendicular* to the other (or sometimes *the perpendicular* to the other). (See Fig. 1.1.5.)

We may write that $\overline{AE}$ is perpendicular to $\overline{DB}$ or more simply that $\overline{AE} \perp \overline{DB}$. It follows that the supplementary angle of a right angle is another right angle.

Multiple perpendicular lines on a many-sided object may be referred to as the object's *perpendiculars*.

The above definition holds for straight lines and rays, *mutatis mutandis*.

A line segment within a triangle that runs from a vertex to an opposite side and is perpendicular to that side is usually referred to an *altitude* of the triangle, although it could be referred to in a general sense as a perpendicular of the triangle.

15. An *acute angle* is one which is less than a right angle. $\angle DAB$ in Fig. 1.1.6 is an acute angle.

16. An *obtuse angle* is one which is greater than a right angle. $\angle EAB$ in Fig. 1.1.6 is an obtuse angle. The supplement of an acute angle is obtuse, and conversely, the supplement of an obtuse angle is acute.

17. When the sum of two angles is a right angle, each is called the *complement* of the other. See Fig. 1.1.6.

---

[18]*Mutatis mutandis* is a Latin phrase meaning "changing [only] those things which need to be changed" or more simply "[only] the necessary changes having been made". Source: http://en.wikipedia.org/wiki/Mutatis_mutandis

1.1. SYMBOLS, LOGIC, AND DEFINITIONS

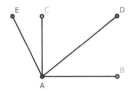

Figure 1.1.6: [Def. 1.17] The angle ∠BAC is a right angle. Since ∠BAC = ∠CAD + ∠DAB, it follows that the angles ∠BAD and ∠DAC are each complements of the other.

Concurrent Lines

18. Three or more straight lines intersecting the same point are called *concurrent* lines. This definition holds for rays and segments, *mutatis mutandis*.

19. The common point through which the rays pass is called the vertex.

The Triangle

20. A triangle is a polygon formed by three segments joined at their endpoints. These three segments are called the *sides* of the triangle. One side in particular may be referred to as the *base* of the triangle for explanatory reasons, but there is no fundamental difference between the properties of a base and the properties of either of the two remaining sides of a triangle. An exception to this is the isosceles triangle where two sides are equal in length: the third side is sometimes referred to as the base.

The formula for the area of a triangle is defined as

$$A = \frac{1}{2}bh$$

where $b =$ the length of a particular side, and $h =$ the length of a perpendicular segment from the base to a vertex.

The area of a triangle is zero if and only if its vertices are collinear.

21. A triangle whose three sides are unequal in length is called *scalene* (the left-hand triangle in Fig. 1.1.7). A triangle with two equal sides is called *isosceles* (the middle triangle in Fig. 1.1.7). When all sides are equal, a triangle is called *equilateral*, (the right-hand triangle in Fig. 1.1.7). When all angles are equal, a triangle is called *equiangular*.

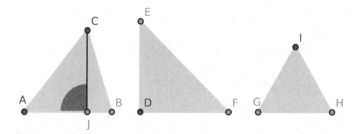

Figure 1.1.7: [Def 1.21] The three types of triangles: *scalene, isosceles, equilateral*.

22. A *right triangle* is a triangle in which one of its angles is a right angle, such as the middle triangle in Fig. 1.1.7. The side which stands opposite the right angle is called the *hypotenuse* of the triangle. (In the middle triangle in Fig. 1.1.7, $\angle EDF$ is a right angle, so side EF is the hypotenuse of the triangle.)

Notice that side EF of a triangle is not written as $\overline{EF}$ despite the fact that EF is also a line segment; since EF is a side of a triangle, we may omit the overline.

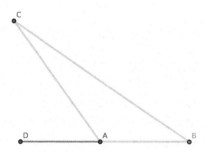

Figure 1.1.8: [Def. 1.23]

23. An *obtuse* triangle is a triangle such that one of its angles is obtuse (such as $\angle CAB$ in $\triangle CAB$, Fig. 1.1.8).

24. An *acute triangle* is a triangle such that each of its angles are acute, such as the left and right triangles in Fig. 1.1.7.

25. An *exterior angle* of a triangle is one which is formed by extending the side of a triangle. In Fig. 1.1.8, $\triangle CAB$ has had side BA extended to the segment $\overline{BD}$ which creates the exterior angle $\angle DAC$.

Every triangle has six exterior angles. Also, each exterior angle is the supplement of the adjacent interior angle. In Fig. 1.1.8, the exterior angle $\angle DAC$ is the supplement of the adjacent interior angle $\angle CAB$.

The Polygon

26. A *rectilinear figure* bounded by three or more line segments can also be referred to as a polygon (see definition 7). For example, a circle is a plane figure but not a polygon, but the triangles in Fig. 1.1.8 are both plane figures and polygons.

27. A polygon is said to be *convex* when it does not have an interior angle greater than 180°.

28. A polygon of four sides is called a *quadrilateral*.

29. A *lozenge*[19] is an equilateral parallelogram whose acute angles are 45 degrees. Sometimes, the restriction to 45 degrees is dropped, and it is required only that two opposite angles are acute and the other two obtuse. The term *rhombus* is commonly used for an equilateral parallelogram[20]; see Fig. 1.1.9.

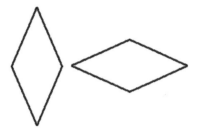

Figure 1.1.9: [Def. 1.29] Two rhombi.

30. A rhombus which has a right angle is called a *square*.

31. A polygon which has five sides is called a *pentagon*; one which has six sides, a *hexagon*, etc.[21]

The Circle

32. A *circle* is a plane figure constructed by connecting all points which are equally distant from a center point. This center point is the *center* of the circle, and the connected points become the *circumference* of the circle.

---

[19]Source: Weisstein, Eric W. "Lozenge." From MathWorld–A Wolfram Web Resource. http://mathworld.wolfram.com/Lozenge.html

[20]Source: Weisstein, Eric W. "Rhombus." From MathWorld--A Wolfram Web Resource. http://mathworld.wolfram.com/Rhombus.html

[21]See also https://en.wikipedia.org/wiki/Polygon

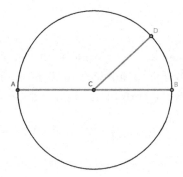

Figure 1.1.10: [Def. 1.32] $\odot C$ is constructed with center C and radius $\overline{CD}$. Notice that $\overline{CA} = \overline{CB} = \overline{CD}$. Also notice that $\overline{AB}$ is a diameter.

33. A *radius* of a circle is any segment constructed from its center to its circumference, such as $\overline{CA}$, $\overline{CB}$, or $\overline{CD}$ in Fig. 1.1.10. Notice that $\overline{CA} = \overline{CB} = \overline{CD}$.

34. A *diameter* of a circle is a segment constructed through the center and terminated at both ends by the circumference, such as $\overline{AB}$ in Fig. 1.1.10.

Other

35. A segment, line, or ray in any figure which divides the area of a regular or symmetrical geometric object into two equal halves is called an *Axis of Symmetry* of the figure (such as AC in the polygon $ABCD$, Fig. 1.1.11).

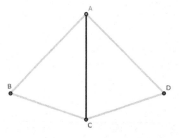

Figure 1.1.11: [Def. 1.35]

Alternatively, if an object is bisected by a segment (or line or ray) such that if for each point on one side of the segment (or line or ray) there exists one point on the other side of the segment (or line or ray) where the distance from each of these points to the segment (or line or ray) is equal, then the segment (or line or ray) is an *Axis of Symmetry*.

36. A segment constructed from any angle of a triangle to the midpoint of the opposite side is called a *median of the triangle*. Each triangle has three medians which are

1.2. POSTULATES

concurrent. The point of intersection of the three medians is called the *centroid* of the triangle.

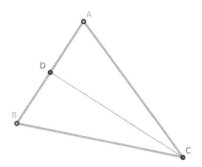

Figure 1.1.12: [Def. 1.36] $\overline{CD}$ is a median of $\triangle ABC$. The triangle has two other medians not shown, and their intersection is the centroid of $\triangle ABC$.

37. A *locus* (plural: loci) is a set of points whose location satisfies or is determined by one or more specified conditions, i.e., 1) every point satisfies a given condition, and 2) every point satisfying it is in that particular locus.[22] For example, a circle is the locus of a point whose distance from the center is equal to its radius.

38. The circumcenter[23] of a triangle is the point where the three perpendicular bisectors of a triangle intersect.

39. The bisectors of the three internal angles of a triangle are concurrent, and their point of intersection is called the **incenter** of the triangle.

Additional definitions will be introduced in [1.5] and subsequent chapters.

1.2 Postulates

We assume the following:

1. A straight line, ray, or segment may be constructed from any one point to any other point. Lines, rays, and segments may be subdivided by points into segments or subsegments which are finite in length.

2. A segment may be extended from any length to a longer segment, a ray, or a straight line.

3. A circle may be constructed from any point (its center) and from any finite length measured from the center (its radius).

---
[22] http://en.wikipedia.org/wiki/Locus_(mathematics)
[23] See also: http://www.mathopenref.com/trianglecircumcenter.html

1.3 Axioms

1.3.1 Algebraic Axioms

Let a, b, c, etc., be real numbers. These axioms require four operations: addition, subtraction, multiplication, and division.

Addition[24] (often signified by the plus symbol "+") is one of the four basic operations of arithmetic, with the others being subtraction, multiplication and division. The addition of two numbers which represent quantities gives us the sum, or total amount of those quantities combined. For example, if $\angle A = \frac{3}{4}\pi$ radians and $\angle B = \frac{1}{4}\pi$ radians, then

$$\angle A + \angle B = \frac{3}{4}\pi + \frac{1}{4}\pi = \left(\frac{3}{4} + \frac{1}{4}\right)\pi = \frac{4}{4}\pi = \pi$$

radians.

Addition for quantities is defined when the quantities under consideration have the same units. For example, a segment with a length of 3 units that is extended by an additional 7 units now has side-length of 10 units; or $3 + 7 = 10$ where each number represents the number of units. However, if a segment with a length of 3 is added to $\angle A = \frac{3}{4}\pi$ radians, the sum is not defined. (In some cases, if the units are removed and the now unit-less numbers are added together, their sum is defined, but the sum may have no bearing on the context of the problem.)

Subtraction represents the operation of removing quantities from a collection of quantities. It is signified by the minus sign ($-$). For example, if $\angle A = \frac{3}{4}\pi$ radians and $\angle B = \frac{1}{4}\pi$ radians, then

$$\angle A - \angle B = \frac{3}{4}\pi - \frac{1}{4}\pi = \left(\frac{3}{4} - \frac{1}{4}\right)\pi = \frac{2}{4}\pi = \frac{1}{2}\pi$$

radians.

Subtraction is a special case of addition where

$$a + (-b) = a - b = c$$

The caveat above concerning units applies to subtraction.

Multiplication (often denoted by the cross symbol "×", by a point "·" or by the absence of symbol): when thinking of multiplication as repeated addition, multiplication is equivalent to adding as many copies of one of them (multiplicand) as the value of

---
[24]Some of this and the following originates from Wikipedia.

1.3. AXIOMS

the other one (multiplier). Normally the multiplier is written first and multiplicand second, though this can vary and sometimes the distinction is not meaningful. As one example,

$$a \times b = a + a + \ldots + a$$

where the product $a \times b$ equals a added to itself a total of b times.

Multiplication is a special case of addition, and so the caveat above concerning units applies to multiplication. The area of a certain geometric objects on the plane (triangles, rectangles, parallelograms, etc.) can be defined by the product of two lengths (base and height, two adjacent side-lengths, etc.). The volume of a certain solid geometric objects (spheres, cubes, etc.) can also be defined by the product of three lengths.

Division: in elementary arithmetic, division (denoted $\div$ or / or by $\frac{a}{b}$ where $b \neq 0$) is an arithmetic operation. Specifically, if b times c equals a, written:

$$a = b \times c$$

where b is not zero, then a divided by b equals c, written:

$$a \div b = c \qquad a/b = c \qquad \frac{a}{b} = c$$

Division is a special case of multiplication where

$$a \cdot \frac{1}{b} = a \div b = \frac{a}{b}$$

Hence, the caveat above concerning units applies to division.

Also, b divides a whenever $a = t \cdot b$ for some integer t; or, $\frac{a}{b} = \frac{tb}{b} = t$ where t is an integer. Note that if b divides a, then we also have that;

1. b is a divisor of a
2. a is a multiple of b
3. a is divisible by b

We use the following as algebraic axioms:[25]

1. The Addition Property: If $a = b$ and $c = d$, then $a + c = b + d$.

---

[25] Sources for these axioms include:
(A) Jurgensen, Brown, Jurgensen. "Geometry." Houghton Mifflin Company, Boston, 1985. ISBN: 0-395-35218-5
(B) Relevant articles in Wikipedia.

2. The Subtraction Property: If $a = b$ and $c = d$, then $a - c = b - d$.

3. Multiplication Property: If $a = b$, then $ca = cb$.

4. Division Property: If $a = b$ and $c \neq 0$, then $\frac{a}{c} = \frac{b}{c}$.

5. Substitution Property: If $a = b$, then either a or b may be substituted for the other in any equation or inequality.

6. Reflexive Property: $a = a$.

7. Symmetric Property: If $a = b$, then $b = a$.

8. Converse Properties of Inequalities:

    (a) If $a \leq b$, then $b \geq a$.

    (b) If $a \geq b$, then $b \leq a$.

9. Transitive Properties of Inequalities:

    (a) If $a \geq b$ and $b \geq c$, then $a \geq c$.

    (b) If $a \leq b$ and $b \leq c$, then $a \leq c$.

    (c) If $a \geq b$ and $b > c$, then $a > c$.

    (d) If $a = b$ and $b > c$, then $a > c$.

10. Inequality Properties of Addition and Subtraction:

    (a) If $a \leq b$, then $a + c \leq b + c$ and $a - c \leq b - c$.

    (b) If $a \geq b$, then $a + c \geq b + c$ and $a - c \geq b - c$.

11. Inequality Properties of Multiplication and Division:

    (a) If $a \geq b$ and $c > 0$, then $ac \geq bc$ and $\frac{a}{c} \geq \frac{b}{c}$.

    (b) If $a \leq b$ and $c > 0$, then $ac \leq bc$ and $\frac{a}{c} \leq \frac{b}{c}$.

    (c) If $a \geq b$ and $c < 0$, then $ac \leq bc$ and $\frac{a}{c} \leq \frac{b}{c}$.

    (d) If $a \leq b$ and $c < 0$, then $ac \geq bc$ and $\frac{a}{c} \geq \frac{b}{c}$.

12. Inequality Property of the Additive Inverse:

    (a) If $a \leq b$, then $-a \geq -b$.

    (b) If $a \geq b$, then $-a \leq -b$.

13. Inequality Property of the Multiplication Inverse (where a and b are either both positive or both negative):

    (a) If $a \leq b$, then $\frac{1}{a} \geq \frac{1}{b}$.

    (b) If $a \geq b$, then $\frac{1}{a} \leq \frac{1}{b}$.

    (c) If $a > (-b)$, then $\frac{1}{b} > \left(-\frac{1}{a}\right)$.

1.3. AXIOMS 31

1.3.2 Congruence Axioms

In geometry, two figures or objects are congruent if they have the same shape and size, or if one has the same shape and size as the mirror image of the other.[26] More formally, two objects are called congruent if and only if one can be transformed into the other using only translations, rotations, or reflections. This means that either object can be re-positioned and reflected (but not resized) so as to coincide precisely with the other object. Therefore two distinct plane figures on a piece of paper are congruent if we can cut them out and then match them up completely. Turning the paper over is permitted.

Examples include:

- Two line segments are congruent if they have the same length.
- Two angles are congruent if they have the same measure.
- Two circles are congruent if they have the same diameter or radius.

If a and b are congruent, we may write $a \cong b$. Three congruence properties with examples:

1. Reflexive Property: $\overline{DE} \cong \overline{DE}$ and $\angle ABC \cong \angle ABC$.
2. Symmetric Property: If $\overline{DE} \cong \overline{FG}$, then $\overline{FG} \cong \overline{DE}$. Also, if $\angle ABC \cong \angle DEF$, then $\angle DEF \cong \angle ABC$.
3. Transitive Property: If $\overline{AB} \cong \overline{CD}$ and $\overline{CD} \cong \overline{EF}$, then $\overline{AB} \cong \overline{EF}$. Also, if $\angle ABC \cong \angle DEF$ and $\angle DEF \cong \angle GHI$, then $\angle ABC \cong \angle GHI$.

The Algebraic and Congruence Axioms give us the Distributive Property:

$$a(b+c) = ab + ac$$

1.3.3 Geometric Axioms

1. Any two objects which can be made to coincide are equal in measure. The placing of one geometrical object on another, such as a line on a line, a triangle on a triangle, or a circle on a circle, etc., is called *superposition*. The superposition employed in geometry is only mental; that is, we conceive of one object being placed on the other. And then, if we can prove that the objects coincide, we infer by the present axiom that they are equal in all respects, including magnitude. Superposition involves the following principle which, without being explicitly stated, Euclid uses frequently: "Any figure may be transferred from one position to another without change in size or form."

---
[26]Taken from https://en.wikipedia.org/wiki/Congruence_(geometry)

2. Two straight lines on a plane cannot enclose a finite area.

3. All right angles are equal to each other.

4. If two lines ($\overleftrightarrow{AB}$, $\overleftrightarrow{CD}$) intersect a third line ($\overleftrightarrow{AC}$) such that the sum of the two interior angles ($\angle BAC + \angle ACD$) on the same side is less than the sum of two right angles, then these lines meet at some finite distance. See Fig. 1.3.1.

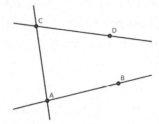

Figure 1.3.1: $\overleftrightarrow{AB}$ and $\overleftrightarrow{CD}$ must eventually meet (intersect) at some finite distance.

The above holds for rays and segments, *mutatis mutandis*.

Playfair's axiom[27] can also be substituted for the above axiom, which states: "In a plane, given a line and a point not on it, at most one line parallel to the given line can be drawn through the point." This axiom was named after the Scottish mathematician John Playfair. His "at most" clause is all that is needed since it can be proven through Euclid's propositions that at least one parallel line exists. This axiom is often written with the phrase, "there is one and only one parallel line".

Axioms which are equivalent to axiom 4 include:

- The sum of the angles in every triangle is 180° (triangle postulate).

- Every triangle can be circumscribed.

- There exists a quadrilateral in which all angles are right angles (that is, a rectangle).

- There exists a pair of straight lines that are at constant distance from each other.

- Two lines that are parallel to the same line are also parallel to each other.

- There is no upper limit to the area of a triangle. (Wallis axiom)[28].

---

[27] https://en.wikipedia.org/wiki/Playfair's_axiom
[28] https://en.wikipedia.org/wiki/Parallel_postulate

1.4 Book I, Propositions 1-26

Proposition 1.1. *CONSTRUCTING AN EQUILATERAL TRIANGLE.*

Given an arbitrary segment, it is possible to construct an equilateral triangle on that segment.

Proof. Suppose we are given segment $\overline{AB}$; we claim that an equilateral triangle can be constructed on $\overline{AB}$.

With A as the center of a circle and $\overline{AB}$ as its radius, we construct the circle $\bigcirc A$ [Postulate 3 from section 1.2]. With B as center and $\overline{AB}$ as radius, we construct the circle $\bigcirc B$, intersecting $\bigcirc A$ at point C.

Construct segments $\overline{CA}$, $\overline{CB}$ [Postulate 1 from section 1.2]. We claim that $\triangle ABC$ is the required equilateral triangle.

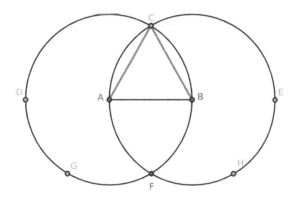

Figure 1.4.1: [1.1]

Because A is the center of the circle $\bigcirc A$, $\overline{AC} = \overline{AB}$ [Def. 1.33]. Since B is the center of the circle $\bigcirc B$, $\overline{AB} = \overline{BC}$. By Axiom 9 from section 1.3.1 (using equalities), we have $\overline{AC} = \overline{AB} = \overline{BC}$.

Since these line segments are the sides of $\triangle ABC$, $\triangle ABC$ is an equilateral triangle [Def. 1.21]. Since $\triangle ABC$ is constructed on segment $\overline{AB}$, we have proven our claim. □

Remark. [1.1]-[1.3] are lemmas to [1.4].

Remark. This proposition may seem strange to readers who are familiar with modern mathematical proofs. Proofs from the 21st century usually show that some nontangible, mathematical object either does or does not exist, while [1.1] describes how to construct an object and then proves that this object is what we intended to construct.

The mathematics of Euclid's day was akin to engineering and construction. If you could prove something existed but not use that knowledge to help construct an object in the real world, you were on track to becoming a full-time *ptochos*[29].

Modern students should draw or construct whatever objects are described in these propositions.

Exam questions.

1. What do we assume in this proposition?

2. What is our claim?

3. What is a finite straight line segment?

4. What is the opposite of finite?

5. What postulates are cited and where are they cited?

6. What axioms are cited and where are they cited?

7. What use is made of the definition of a circle? What is a circle?

8. What is an equilateral triangle?

Exercises.

Exercises #2-5 should be attempted after the student has completed Chapter 1.

1. If the segments $\overline{AF}$ and $\overline{BF}$ are constructed, prove that the figure $\square ACBF$ is a rhombus. [See the final chapter for a solution.]

2. If $\overline{CF}$ is constructed and $\overline{AB}$ is extended to the circumferences of the circles (at points D and E), prove that the triangles $\triangle CDF$ and $\triangle CEF$ are equilateral. [See the final chapter for a solution.]

3. If $\overline{CA}$ and $\overline{CB}$ are extended to intersect the circumferences at points G and H, prove that the points G, F, H are collinear and that the triangle $\triangle GCH$ is equilateral.

4. Construct $\overline{CF}$ and prove that $\left(\overline{CF}\right)^2 = 3 \cdot \left(\overline{AB}\right)^2$.

5. Construct a circle in the space ACB bounded by the segment $\overline{AB}$ and the partial circumferences of the two circles.

---

[29]https://en.wikipedia.org/wiki/Begging

1.4. BOOK I, PROPOSITIONS 1-26

Proposition 1.2. *CONSTRUCTING A LINE SEGMENT EQUAL IN LENGTH TO AN ARBITRARY LINE SEGMENT.*

Given an arbitrary point and an arbitrary segment, it is possible to construct a segment with:

(1) one endpoint being the previously given point

(2) a length equal to that of the arbitrary segment.

Proof. Let A be an arbitrary point on the plane, and let $\overline{BC}$ be an arbitrary segment. Our claim is stated above.

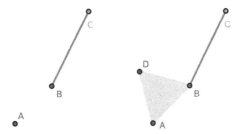

Figure 1.4.2: [1.2] at the beginning of the proof (left), and then partially constructed (right)

Construct $\overline{AB}$, and on $\overline{AB}$ construct the equilateral triangle $\triangle ABD$ [1.1].

With B as the center and $\overline{BC}$ as the radius, construct $\odot B$. Extend $\overline{DB}$ to intersect the circle $\odot B$ at E [Postulate 2 from section 1.2]. With D as the center and $\overline{DE}$ as radius, construct $\odot D$. Extend $\overline{DA}$ to meet $\odot B$ at F. □

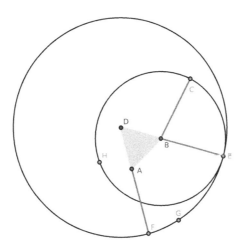

Figure 1.4.3: [1.2] fully constructed

Proof. Clearly, $\overline{AF}$ has A as one of its endpoints (claim 1). If we can show that $\overline{AF} = \overline{BC}$, we will have proven our claim.

Since $\overline{DE}$ and $\overline{DF}$ are radii of $\odot D$, $\overline{DF} = \overline{DE}$ [Def. 1.32]. Because $\triangle DAB$ is an equilateral triangle, $\overline{DA} = \overline{DB}$ [Def. 1.21]. By [Axiom 2 from section 1.3.1], we find that

$$\overline{DF} - \overline{DA} = \overline{DE} - \overline{DB}$$

But $\overline{DF} - \overline{DA} = \overline{AF}$ and $\overline{DE} - \overline{DB} = \overline{BE}$. By [Axiom 5 from section 1.3.1], we find that $\overline{AF} = \overline{BE}$.

Since $\overline{BC}$ and $\overline{BE}$ are radii at $\odot B$, $\overline{BE} = \overline{BC}$. By Axiom 9 from section 1.3.1 (using equalities), we have $\overline{AF} = \overline{BC}$ (claim 2), which completes the proof. $\square$

Exercises.

1. Prove [1.2] when A is a point on $\overline{BC}$. [See the final chapter for a solution.]

1.4. BOOK I, PROPOSITIONS 1-26

Proposition 1.3. *SUBDIVIDING A LINE SEGMENT.*

Given two arbitrary, unequal segments, it is possible to subdivide the larger segment such that one of its two sub-segments is equal in length to the smaller segment.

Proof. Construct segments $\overline{AB}$ and $\overline{CG}$ such that $\overline{CG} < \overline{AB}$. We claim that $\overline{AB}$ may be subdivided into segments $\overline{AE}$ and $\overline{EB}$ where $\overline{AE} = \overline{CG}$.

From A, construct $\overline{AD}$ such that $\overline{AD} = \overline{CG}$ [1.2]. With A as the center and $\overline{AD}$ as radius, construct $\bigcirc A$ [Postulate 1.3] which intersects $\overline{AB}$ at E.

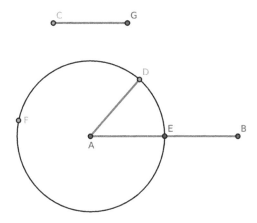

Figure 1.4.4: [1.3]

Because A is the center of $\bigcirc A$, $\overline{AE} = \overline{AD}$ [Def. 1.32]. Since $\overline{AD} = \overline{CG}$ by construction, by Axiom 9 from section 1.3.1 (using equalities), $\overline{AE} = \overline{CG}$, which proves our claim. □

Corollary. *1.3.1. Given arbitrary segments and a ray, it is possible to cut the ray such that the resulting segment is equal in length to the arbitrary segment.*

Exam questions.

1. What previous problem is employed in the solution of this?

2. What axiom is employed in the demonstration?

3. Demonstrate how to extend the shorter of the two given segments until the whole extended segment is equal in length to the longer segment.

Exercises.

1. Prove [Cor. 1.3.1].

Proposition 1.4. *THE "SIDE-ANGLE-SIDE" THEOREM FOR THE CONGRUENCE OF TRIANGLES.*

If two pairs of sides in two triangles is respectively equal in length, and if the corresponding interior angles are equal in measure, then the triangles are congruent.

Proof. If $\triangle ABC$ and $\triangle DEF$ exist such that $\overline{AB} = \overline{DE}$, $\overline{AC} = \overline{DF}$, and $\angle BAC = \angle EDF$, then $\triangle ABC \cong \triangle DEF$.

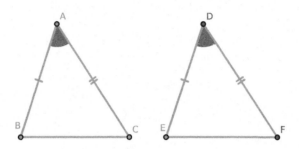

Figure 1.4.5: [1.4]

Recall that superposition allows us to move one object top of another without distorting its shape or measure. If $\triangle ABC$ is positioned[30] on $\triangle DEF$ such that the point A is positioned on the point D and side AB is positioned on side DE, then the point B coincides with the point E because $AB = DE$.

Since AB coincides with DE, the side AC also coincides with side DF because $\angle BAC = \angle EDF$. Since $AC = DF$, C coincides with F.

Because B coincides with E, the base BC of $\triangle ABC$ coincides with the base EF of $\triangle DEF$; it follows that $BC = EF$.

Hence all sides and angles of one triangle are equal with their corresponding sides and angles in the other triangle. We conclude that $\triangle ABC \cong \triangle DEF$. □

Remark. Euclid's Elements contains three propositions on the congruence of triangles: [1.4] side-angle-side (SAS), [1.8] side-side-side (SSS), and [1.26] angle-angle-side (AAS or SAA) and angle-side-angle (ASA).

Exam questions.

1. What is meant by superposition?

2. How many parts make up a triangle? (Ans. 6, three sides and three angles.)

---

[30] We may write "positioned" instead of "superpositioned" with the understanding that the words are synonymous in context.

3. When it is required to prove that two triangles are congruent, how many parts of one must be given equal to the corresponding parts of the other? (Ans. In general, any three except the three angles. This will be established in [1.8] and [1.26], both of which use [1.4].)

Exercises.

1. Prove that the line which bisects the vertical angle of an isosceles triangle also bisects the base perpendicularly. [See the final chapter for a solution.]

2. If two adjacent sides of a quadrilateral are equal in length and the diagonal bisects the angle between them, prove that their remaining sides are also equal in length. [See the final chapter for a solution.]

3. If two segments stand perpendicularly to each other and if each bisects the other, prove that any point on either segment is equally distant from the endpoints of the other segment. [See the final chapter for a solution.]

4. If equilateral triangles are constructed on the sides of any triangle, prove that the distances between the vertices of the original triangle and the opposite vertices of the equilateral triangles are equal. (This may be proven after studying [1.32].)

Proposition 1.5. *ISOSCELES TRIANGLES I.*

If a triangle is isosceles, then:

(1) if the sides of the triangle other than the base are extended, the angles under the base are equal in measure

(2) the angles at the base are equal in measure.

Proof. Construct $\triangle ABC$ such that sides $AB = AC$ and denote side BC as the triangle's base. Extend $\overline{AB}$ to $\overline{AD}$ and $\overline{AC}$ to $\overline{AE}$ such that $\overline{CE} \geq \overline{BD}$. We claim that:

(1) $\angle DBC = \angle ECB$

(2) $\angle ABC = \angle ACB$

We will prove each claim separately. Claim 1: $\angle DBC = \angle ECB$

Let F be a point on $\overline{BD}$ other than B or D. On $\overline{CE}$, choose point G such that $\overline{CG} = \overline{BF}$ [1.3]. (Since $\overline{CE} \geq \overline{BD}$, such a point exists and is not an endpoint.) Construct $\overline{BG}$ and $\overline{CF}$ [two applications of Postulate 1 from section 1.2].

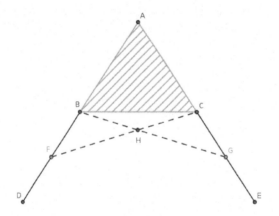

Figure 1.4.6: [1.5]

Because $\overline{AF} = \overline{AG}$ by construction and $\overline{AB} = \overline{AC}$ by hypothesis, it follows that sides AF and AC in $\triangle FAC$ are respectively equal in length to sides AG and AB in $\triangle GAB$. Also, the angle $\angle BAC$ is the interior angle to both pairs of sides in each triangle. By [1.4], $\triangle FAC \cong \triangle GAB$; this implies that $\angle AFC = \angle AGB$ and $\overline{BG} = \overline{CF}$.

Consider $\triangle FBC$, $\triangle GCB$: since $\overline{BF} = \overline{CG}$, $\overline{CF} = \overline{BG}$, and $\angle AFC = \angle AGB$, by [1.4] $\triangle FBC \cong \triangle GCB$. This implies that $\angle FBC = \angle GCB$, which are the angles under the base of $\triangle ABC$; or,

$$\angle DBC = \angle FBC = \angle GCB = \angle ECB$$

This proves claim 1.

1.4. BOOK I, PROPOSITIONS 1-26

Claim 2: $\angle ABC = \angle ACB$

Since $\triangle FBC \cong \triangle GCB$, we have $\angle FCB = \angle GBC$. By the above[31], $\angle FCA = \angle GBA$. Notice that:

$$\angle FCA = \angle GBA$$
$$\angle FCB + \angle ACB = \angle GBC + \angle ABC$$
$$\angle FCB + \angle ACB = \angle FCB + \angle ABC$$
$$\angle ACB = \angle ABC$$

This proves claim 2 and completes the proof. □

Remark. The difficulty which beginners may have with this proposition is due to the fact that $\triangle ACF$, $\triangle ABG$ overlap. A teacher or tutor should graph these triangles separately and point out the corresponding parts: $AF = AG$, $AC = AB$, and $\angle FAC = \angle GAB$. By [1.4], it follows that $\angle ACF = \angle ABG$, $\angle AFC = \angle AGB$.

Corollary. *1.5.1. A triangle is equilateral if and only if it is equiangular.*

Exercises.

1. Prove that the angles at the base are equal without extending the sides.

2. Prove that $\overleftrightarrow{AH}$ is an Axis of Symmetry of $\triangle ABC$. [See the final chapter for a solution.]

3. Prove that each diagonal of a rhombus is an Axis of Symmetry of the rhombus.

4. Take the midpoint on each side of an equilateral triangle; the segments joining them form a second equilateral triangle. [See the final chapter for a solution.]

5. Prove [Cor. 1.5.1].

> "Detection is, or ought to be, an exact science, and should be treated in the same cold unemotional manner. You have attempted to tinge it with romanticism, which produces the same effect as if you worked a love-story into the fifth proposition of Euclid."
> - Sir Arthur Conan Doyle, "The Sign of Four"[a]
>
> [a]https://gutenberg.org/ebooks/2097

---

[31]"By the above" means "This is a result from earlier in the proof. You may need to reread the proof to find it, but it's there."

Proposition 1.6. *ISOSCELES TRIANGLES II.*

If a given triangle has two equal angles, then the sides opposite the two angles are equal in length (i.e., the triangle is isosceles).

Proof. Construct $\triangle ABC$ such that $\angle ABC = \angle ACB$. In order to prove our claim that $\overline{AB} = \overline{AC}$, we will use a proof by contradiction[32][33].

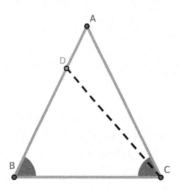

Figure 1.4.7: [1.6]

Without loss of generality[34], suppose side $AB > AC$. On AB, construct a point D such that $\overline{BD} = \overline{CA}$ [1.3] and construct $\overline{CD}$. Notice that $\triangle ACD > 0$ (i.e., the area of $\triangle ACD$ is greater than 0); otherwise, we would not have $\triangle ACD$.

Consider $\triangle DBC$ and $\triangle ACB$: since $DB = AC$, $\angle DBC = \angle ACB$, and each triangle contains the side BC, by [1.4] $\triangle DBC \cong \triangle ACB$. It follows that $\triangle ACB = \triangle DBC$.

But $\triangle ACB = \triangle ACD \oplus \triangle DBC$, and so it follows that $\triangle ACD = 0$ (i.e., the area of $\triangle ACD$ equals 0). But above we showed that $\triangle ACD > 0$. By showing that $\triangle ACD = 0$ and $\triangle ACD > 0$, we obtain a contradiction.

Specifically, we assumed that $AB > AC$ and obtained a contradiction. If we instead assume that $AC > AB$, we would obtain the same contradiction (this is what we mean by "without loss of generality"; we had two ways to begin the proof, and either way would have obtained the same result). Since $AB > AC$ and $AC > AB$ each produce a contradiction, we must have $AB = AC$, which proves our claim. □

---

[32]Mathematics rigidly follows the laws of Western logic, which means that contradictions are always a sign of error. If we attempt to prove a proposition and obtain a contradiction, then, because a proposition is either true or false, the proposition must be false.
See also: https://en.wikipedia.org/wiki/Contradiction

[33]This proof by contradiction will show that the statement "A triangle with two equal angles has unequal opposite sides" is false.
See also: https://en.wikipedia.org/wiki/Proof_by_contradiction

[34]This term is used before an assumption in a proof which narrows the premise to some special case; it is implied that either the proof for that case can be easily applied to all others or that all other cases are equivalent. Thus, given a proof of the conclusion in the special case, it is trivial to adapt it to prove the conclusion in all other cases. It is usually abbreviated as *wlog*.
See also: http://en.wikipedia.org/wiki/Without_loss_of_generality

1.4. BOOK I, PROPOSITIONS 1-26

Corollary. *1.6.1. Together, [1.5] and [1.6] state that a triangle is isosceles if and only if[35] the angles at its base are equal.*

Exam questions.

1. What is the hypothesis in this proposition?
2. What proposition is this the converse of?
3. What is the inverse of this proposition?
4. What is the inverse of [1.5]?
5. What is meant by a proof by contradiction?
6. How does Euclid generally prove converse propositions?
7. What false assumption is made in order to prove the proposition?
8. What does this false assumption lead to?

Exercises.

1. Prove [Cor. 1.6.1].

---

[35] Often abbreviated as "iff".
See also: https://en.wikipedia.org/wiki/If_and_only_if

Proposition 1.7. DISTINCT TRIANGLES.

If we construct two distinct triangles such that two sides are equal in length, then the third side of each triangle will be unequal in length.

Proof. Construct distinct triangles $\triangle ADB$, $\triangle ACB$ which share the base AB. Suppose that $AC = AD$. We claim that $BC \neq BD$.

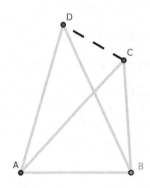

Figure 1.4.8: [1.7], case 1

The vertex of the second triangle may be either within or outside the first triangle.

Case 1: *Vertex outside of the other triangle.*

Let the vertex of each triangle lie outside the interior of the other triangle (i.e., D does not lie inside $\triangle ACB$ and C does not lie inside $\triangle ADB$). Construct $\overline{CD}$. Because $AD = AC$ by hypothesis, $\triangle ACD$ is isosceles. By [Cor. 1.6.1], $\angle ACD = \angle ADC$.

Since $\angle ADC = \angle ADB + \angle BDC$ and $\angle ADB > 0$, it follows that $\angle ADC > \angle BDC$ [1.3.1 Axiom 12]. Because $\angle ACD = \angle ADC$, we also have that $\angle ACD > \angle BDC$. Since $\angle BCD = \angle BCA + \angle ACD$, we also have that $\angle BCD > \angle BDC$.

Consider $\triangle BDC$. Since $\angle BCD > \angle BDC$, by [Cor. 1.6.1] we find that $BD \neq BC$, which proves our claim.

Case 2: *Vertex inside the other triangle.*

Wlog, let the vertex of the triangle $\triangle ADB$ be located within the interior of $\triangle ACB$. Extend side AC to segment $\overline{AE}$ and side AD to segment $\overline{AF}$. Construct $\overline{CD}$. Because $AC = AD$ by hypothesis, the triangle $\triangle ACD$ is isosceles; by [1.5] $\angle ECD = \angle FDC$.

1.4. BOOK I, PROPOSITIONS 1-26

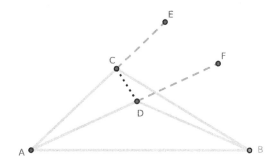

Figure 1.4.9: [1.7], case 2

Since $\angle ECD = \angle ECB + \angle BCD$, it follows that $\angle ECD > \angle BCD$ [1.3.1 Axiom 12]. Because $\angle ECD = \angle FDC$, we also have that $\angle FDC > \angle BCD$. Since $\angle BDC = \angle BDF + \angle FDC$, we find that $\angle BDC > \angle BCD$.

Consider $\triangle BDC$. Since $\angle BDC > \angle BCD$, by [Cor. 1.6.1] we find that $BD \neq BC$, which proves our claim and completes the proof. □

Corollary. *1.7.1. Two triangles are distinct whenever no side of one triangle is equal in length to any side of the other triangle.*

Exercises.

1. Prove [Cor. 1.7.1].

Proposition 1.8. *THE "SIDE-SIDE-SIDE" THEOREM FOR THE CONGRUENCE OF TRIANGLES.*

If all three pairs of sides of two triangles are respectively equal in length, then the triangles are congruent.

Proof. Suppose $\triangle ABC$ and $\triangle DEF$ are triangles where sides $AB = DE$, $AC = DF$, and $BC = EF$ (where BC and EF are the bases of those triangles). We claim that $\triangle ABC \cong \triangle DEF$.

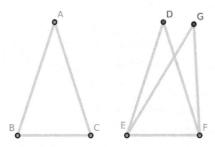

Figure 1.4.10: [1.8]

Let $\triangle ABC$ be positioned on $\triangle DEF$ such that point B coincides with point E and the side BC coincides with the side EF. Because $BC = EF$, the point C coincides with point F. If the vertex A falls on the same side of EF as vertex D, then the point A must coincide with D.

If this were not true, then A must have a different location: call this point G. Our hypothesis then states $EG = AB$ and $AB = ED$. By [1.3.1 Axiom 8], $EG = ED$. Similarly, $FG = FD$. However, by [1.7], $FG \neq FD$, a contradiction.

Hence the point A must coincide with point D, and so the three angles of one triangle are respectively equal to the three angles of the other (specifically, $\angle ABC = \angle DEF$, $\angle BAC = \angle EDF$, and $\angle BCA = \angle EFD$). Therefore, $\triangle ABC \cong \triangle DEF$. $\square$

Exam questions.

1. What use is made of [1.7]? (Ans: As a lemma to [1.8].)

2. Can [1.7] and [1.8] be combined into a single proposition? If so, how?

1.4. BOOK I, PROPOSITIONS 1-26

Proposition 1.9. *BISECTING A RECTILINEAR ANGLE.*

It is possible to bisect an angle.

Proof. Construct $\angle BAC$, point D on $\overrightarrow{AB}$, and point E on $\overrightarrow{AC}$ such that $\overline{AE} = \overline{AD}$ [1.3]. Construct $\overline{DE}$; also construct the equilateral triangle $\triangle DEF$ [1.1] such that F stands on the other side of $\overline{DE}$ than A. Construct $\overline{AF}$. We claim that $\overline{AF}$ bisects $\angle BAC$.

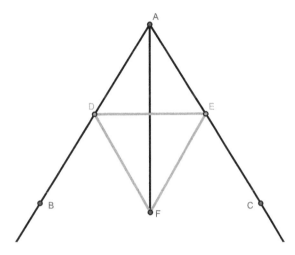

Figure 1.4.11: [1.9]

Consider $\triangle DAF$, $\triangle EAF$: each shares $\overline{AF}$, $\overline{AD} = \overline{AE}$ by construction, and $\overline{DF} = \overline{EF}$ by construction. By [1.8], $\triangle DAF \cong \triangle EAF$, and so $\angle DAF = \angle EAF$. Notice that

$$\angle BAC = \angle DAF + \angle EAF$$
$$= 2 \cdot \angle DAF$$
$$\implies$$
$$\frac{1}{2} \cdot \angle BAC = \angle DAF$$

Or, $\angle BAC$ is bisected by $\overline{AF}$, which completes the proof. $\square$

Corollary. *1.9.1.* If $\overline{AF}$ is extended to the line $\overleftrightarrow{AF}$, then $\overleftrightarrow{AF}$ is the Axis of Symmetry of the $\triangle AED$, $\triangle DEF$, figure $BDAEC$, and segment $\overline{DE}$.

Corollary. *1.9.2.* In [1.9], AB and AC may be constructed as lines, rays, or segments of appropriate length with point A as the vertex, mutatis mutandis.

Exam questions.

1. Why does Euclid construct the equilateral triangle on the side opposite of A?

2. If the equilateral triangle were constructed on the other side of DE, in what case would the construction fail?

Exercises.

1. Prove [1.9] without using [1.8]. (Hint: use [1.5, #2].)

2. Prove that $\overline{AF} \perp \overline{DE}$. (Hint: use [1.5, #2].) [See the final chapter for a solution.]

3. Prove that any point on $\overline{AF}$ is equally distant from the points D and E. [See the final chapter for a solution.]

4. Prove [Cor. 1.9.1].

5. Prove [Cor. 1.9.2].

1.4. BOOK I, PROPOSITIONS 1-26

Proposition 1.10. *BISECTING A SEGMENT.*

It is possible to bisect a segment of arbitrary length (i.e., it is possible to locate the midpoint of a segment).

Proof. Construct $\overline{AB}$; we claim that the segment $\overline{AB}$ can be bisected.

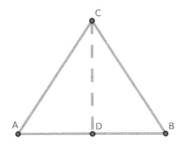

Figure 1.4.12: [1.10]

Construct the equilateral triangle $\triangle ABC$ with AB as its base [1.1]. Bisect $\angle ACB$ by constructing the segment $\overline{CD}$ [1.9] which intersects $\overline{AB}$ at D. Clearly, $\overline{AB} = \overline{AD} \oplus \overline{DB}$. We claim that $\overline{AB} = 2 \cdot \overline{AD} = 2 \cdot \overline{DB}$ (which is equivalent to stating that $\overline{AB}$ is bisected at D).

Consider $\triangle ACD$, $\triangle BCD$: $AC = BC$ (since each are sides of the equilateral triangle $\triangle ACB$); each triangle shares side CD; $\angle ACD = \angle BCD$ by construction. By [1.4], $\triangle ACD \cong \triangle BCD$, and so $\overline{AD} = \overline{DB}$. Therefore,

$$\overline{AB} = \overline{AD} + \overline{DB} = 2 \cdot \overline{AD} = 2 \cdot \overline{DB}$$

which proves our claim. □

Exercises.

1. Bisect a segment by constructing two circles. [See the final chapter for a solution.]

2. Extend $\overline{CD}$ to $\overleftrightarrow{CD}$. Prove that every point equally distant from the points A and B are points on $\overleftrightarrow{CD}$. [See the final chapter for a solution.]

Proposition 1.11. *CONSTRUCTING A PERPENDICULAR SEGMENT I.*

It is possible to construct a segment at a right angle to a given line from any point on the line.

Proof. Construct $\overleftrightarrow{AB}$ containing point C. On $\overrightarrow{CA}$, choose any point D; on $\overrightarrow{CB}$, choose E such that $\overline{CE} = \overline{CD}$ [1.3]. Construct the equilateral triangle $\triangle DFE$ on $\overline{DE}$ [1.1] and construct $\overline{CF}$. We claim that $\overleftrightarrow{AB} \perp \overline{CF}$.

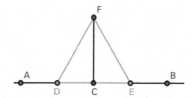

Figure 1.4.13: [1.11]

Consider $\triangle DCF$, $\triangle ECF$: each shares side CF, $CD = CE$ by construction, and $DF = EF$ since $\triangle DFE$ is equilateral. By [1.8] $\triangle DCF \cong \triangle ECF$, and so $\angle DCF = \angle ECF$. Since these are adjacent angles, [Def. 1.13] states that each of these angles is a right angle, which proves our claim. □

Corollary. *1.11.1. [1.11] holds when AB is a segment or ray and/or when CF is a straight line or a ray, mutatis mutandis.*

Exercises.

1. Prove that the diagonals of a rhombus bisect each other perpendicularly. [See the final chapter for a solution.]

2. Prove [1.11] without using [1.8].

3. Find a point on a given line that is equally distant from two given points. [See the final chapter for a solution.]

4. Find a point on a given line such that if it is joined to two given points on opposite sides of the line, then the angle formed by the connecting segment is bisected by the given line. (Hint: similar to the proof of #3.)

5. Find a point that is equidistant from three given points. (Hint: you are looking for the circumcenter of the triangle.)

6. Prove [Cor. 1.11.1].

1.4. BOOK I, PROPOSITIONS 1-26

Proposition 1.12. *CONSTRUCTING A PERPENDICULAR SEGMENT II.*

Given an arbitrary line and an arbitrary point not on the line, we may construct a perpendicular segment from the point to the line.

Proof. Construct $\overleftrightarrow{AB}$ and C such that C is not on $\overleftrightarrow{AB}$. We wish to construct $\overline{CH}$ such that H is on $\overleftrightarrow{AB}$ and $\overline{CH} \perp \overleftrightarrow{AB}$.

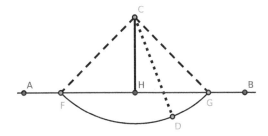

Figure 1.4.14: [1.12]

Take any point D from the opposite side of $\overleftrightarrow{AB}$ to C. Construct the circle $\odot C$ with $\overline{CD}$ as its radius [Postulate 1.3] where $\odot C$ intersects $\overleftrightarrow{AB}$ at the points F and G. Bisect $\overline{FG}$ at H [1.10] and construct $\overline{CH}$ [Postulate 1.1]. We claim that $\overline{CH} \perp \overleftrightarrow{AB}$.

Construct $\triangle CFG$, and consider $\triangle FHC$ and $\triangle GHC$: $\overline{FH} = \overline{GH}$ by construction; the triangles share $\overline{HC}$; $\overline{CF} = \overline{CG}$ since each are radii of $\odot C$ [Def. 1.32]. By [1.8], $\triangle FHC \cong \triangle GHC$, and so $\angle CHF = \angle CHG$. Since these are adjacent angles, [Def. 1.13] states that each angle is a right angle, which proves our claim. $\square$

Corollary. *1.12.1. [1.12] holds when CH and/or AB are replaced by rays, mutatis mutandis.*

Exercises.

1. Prove that circle $\odot C$ cannot meet $\overleftrightarrow{AB}$ at more than two points. [See the final chapter for a solution.]

2. Prove [Cor. 1.12.1].

CHAPTER 1. ANGLES, PARALLEL LINES, PARALLELOGRAMS

Proposition 1.13. *ANGLES AT INTERSECTIONS OF STRAIGHT LINES.*

If a line intersects another line, the lines either stand at right angles or at two angles whose sum equals two right angles.

Proof. If the line $\overleftrightarrow{AB}$ intersects the line $\overleftrightarrow{CD}$ at B, we claim that either $\angle ABC$ and $\angle ABD$ are right angles or that $\angle ABC + \angle ABD$ equals two right angles.

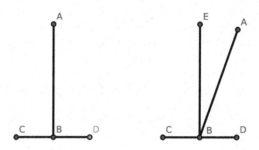

Figure 1.4.15: [1.13] (α) on left, (β) on right

If $\overleftrightarrow{AB} \perp \overleftrightarrow{CD}$ as in Fig. 1.4.15(α), then $\angle ABC$ and $\angle ABD$ are right angles.

Otherwise, $\angle ABC$ and $\angle ABD$ are not right angles as in Fig. 1.4.15(β); construct $\overleftrightarrow{BE} \perp \overleftrightarrow{CD}$ [1.11]. Notice that $\angle ABC = \angle CBE + \angle EBA$ [Def. 1.11]. Adding $\angle ABD$ to each side of this equality, we obtain that

$$\angle ABC + \angle ABD = \angle CBE + \angle EBA + \angle ABD$$

Similarly, we find that

$$\angle CBE + \angle EBA + \angle ABD = \angle CBE + \angle EBD$$

By [1.3.1 Axiom 8], we find that

$$\angle ABC + \angle ABD = \angle CBE + \angle EBD$$

Since $\angle CBE$ and $\angle EBD$ are right angles, $\angle ABC + \angle ABD$ equals the sum of two right angles, which proves our claim. □

An alternate proof:

Proof. Denote $\angle EBA$ by θ. Notice that:

$$\angle CBA = \text{right angle} + \theta$$
$$\angle ABD = \text{right angle} - \theta$$
$$\implies$$
$$\angle CBA + \angle ABD = \text{right angle}$$

□

Corollary. *1.13.1. The above proposition holds when the straight lines are replaced by segments and/or rays, mutatis mutandis.*

Corollary. *1.13.2. The sum of two supplemental angles equals two right angles.*

Corollary. *1.13.3. Two distinct straight lines cannot share a common segment.*

Corollary. *1.13.4. The bisector of any angle bisects the corresponding re-entrant angle.*

Corollary. *1.13.5. The bisectors of two supplemental angles are at right angles to each other.*

Corollary. *1.13.6. The angle $\angle EBA = \frac{1}{2} \cdot (\angle CBA - \angle ABD)$.*

Exercises.

1. Prove [Cor. 1.13.1].
2. Prove [Cor. 1.13.2].
3. Prove [Cor. 1.13.3].
4. Prove [Cor. 1.13.4].
5. Prove [Cor. 1.13.5].
6. Prove [Cor. 1.13.6].

Proposition 1.14. *RAYS TO STRAIGHT LINES.*

If at the endpoint of a ray there exists two other rays constructed on opposite sides of the first ray such that the sum of their adjacent angles is equal to two right angles, then these two rays form one line.

Proof. Construct $\overrightarrow{BA}$. On opposite sides of $\overrightarrow{BA}$, construct $\overrightarrow{BC}$ and $\overrightarrow{BD}$ such that the sum of their adjacent angles, $\angle CBA + \angle ABD$, equals two right angles. We claim that $\overrightarrow{BC} \oplus \overrightarrow{BD} = \overleftrightarrow{CD}$.

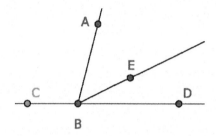

Figure 1.4.16: [1.14]

Suppose instead that $\overrightarrow{BC} \oplus \overrightarrow{BE} = \overleftrightarrow{CE}$ and $\angle EBD > 0$. Since $\overleftrightarrow{CE}$ is a line and $\overrightarrow{BA}$ stands on it, the sum $\angle CBA + \angle ABE$ equals two right angles [1.13]. Also by hypothesis, the sum $\angle CBA + \angle ABD$ equals two right angles. Therefore,

$$\begin{aligned} \angle CBA + \angle ABE &= \angle CBA + \angle ABD \\ \angle ABE &= \angle ABD \\ \angle ABE &= \angle ABE + \angle EBD \\ 0 &= \angle EBD \end{aligned}$$

Since $\angle EBD = 0$ and $\angle EBD > 0$, we have a contradiction. Hence, $\overrightarrow{BC} \oplus \overrightarrow{BD} = \overleftrightarrow{CD}$, which proves our claim. □

Corollary. *1.14.1. The above result holds for segments, mutatis mutandis.*

Exercises.

1. Prove [Cor. 1.14.1].

1.4. BOOK I, PROPOSITIONS 1-26

Proposition 1.15. *OPPOSITE ANGLES ARE EQUAL.*

If two lines intersect at a point, then their opposite angles are equal.

Proof. Suppose $\overleftrightarrow{AB}$ and $\overleftrightarrow{CD}$ intersect at E. We claim that $\angle AEC = \angle DEB$ and $\angle BEC = \angle DEA$.

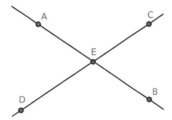

Figure 1.4.17: [1.15]

Because $\overleftrightarrow{AB}$ intersects $\overleftrightarrow{CD}$ at E, the sum $\angle DEA + \angle AEC$ equals two right angles [1.13]. Similarly, because $\overleftrightarrow{CD}$ intersects $\overleftrightarrow{AB}$ at E, the sum $\angle AEC + \angle BEC$ also equals two right angles. Therefore,

$$\angle AEC + \angle DEA = \angle AEC + \angle BEC$$
$$\angle DEA = \angle BEC$$

Similarly, we can also show that $\angle AEC = \angle DEB$, which proves our claim. □

An alternate proof:

Proof. Because opposite angles share a common supplement, they are equal. □

Corollary. *1.15.1. [1.15] holds when either one or both of the two straight lines are replaced either by segments or by rays, mutatis mutandis.*

Exam questions for [1.13]–[1.15].

1. What problem is required in Euclid's proof of [1.13]?

2. What theorem? (Ans. No theorem, only the axioms.)

3. If two lines intersect, how many pairs of supplemental angles do they make?

4. What is the relationship between [1.13] and [1.14]?

5. What three lines in [1.14] are concurrent?

6. State the converse of Proposition [1.15] and prove it.

7. What is the subject of [1.13], [1.14], [1.15]? (Ans. Angles at a point.)

Exercises.

1. Prove [Cor. 1.15.1].

1.4. BOOK I, PROPOSITIONS 1-26

Proposition 1.16. *THE EXTERIOR ANGLE OF A TRIANGLE IS GREATER THAN EITHER OF THE NON-ADJACENT INTERIOR ANGLES.*

If any side of a triangle is extended, the resulting exterior angle is greater than either of the non-adjacent interior angles.

Proof. Construct $\triangle ABC$. Wlog, we extend side BC to $\overline{BD}$. We claim that the exterior angle $\angle ACD$ is greater than either of the interior non-adjacent angles $\angle ABC$, $\angle BAC$.

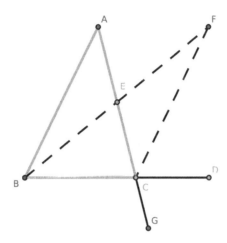

Figure 1.4.18: [1.16]

Bisect AC at E [1.10] and construct $\overline{BE}$ [Postulate 1.1]. Extend $\overline{BE}$ to $\overline{BF}$ such that $\overline{BE} = \overline{EF}$ [1.3]. Also construct $\overline{CF}$.

Consider $\triangle CEF$ and $\triangle AEB$: $\overline{CE} = \overline{EA}$ by construction, $\overline{BE} = \overline{EF}$ by construction, and $\angle CEF = \angle AEB$ [1.15]. By [1.4], $\triangle CEF \cong \triangle AEB$, and so $\angle ECF = \angle EAB$. Since $\angle ACD = \angle ECF + \angle FCD$ and $\angle EAB = \angle BAC$,

$$\angle ACD = \angle EAB + \angle FCD = \angle BAC + \angle FCD$$

It follows that $\angle ACD > \angle BAC$.

Similarly, if BC is bisected, it can be shown that $\angle ACD > \angle ABC$, which completes the proof. □

Corollary. 1.16.1. *The sum of the three interior angles of the triangle $\triangle BCF$ is equal to the sum of the three interior angles of the triangle $\triangle ABC$.*

Corollary. 1.16.2. *The area of $\triangle BCF$ is equal to the area of $\triangle ABC$, which we will write as $\triangle BCF = \triangle ABC$.*

Corollary. *1.16.3. If sides BA and CF are extended to lines, they cannot meet at any finite distance. For, if they met at some point X, then the triangle $\triangle CAX$ would have an exterior angle $\angle BAC$ equal to the interior angle $\angle ACX$.*

Exercise.

1. Prove [Cor. 1.16.1].

2. Prove [Cor. 1.16.2].

3. Prove [Cor. 1.16.3] using a proof by contradiction.

1.4. BOOK I, PROPOSITIONS 1-26

Proposition 1.17. *THE SUM OF TWO INTERIOR ANGLES OF A TRIANGLE.*

The sum of two interior angles of a triangle is less than the sum of two right angles.

Proof. We claim that the sum of any two interior angles of $\triangle ABC$ is less than the sum of two right angles.

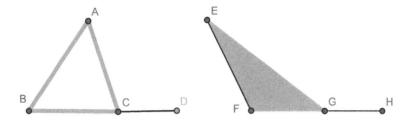

Figure 1.4.19: [1.17]

Wlog, choose $\angle ABC$ and $\angle BAC$ and extend side BC to $\overline{BD}$. By [1.16], $\angle ACD > \angle ABC$. To each, add the angle $\angle ACB$:

$$\angle ACD + \angle ACB \quad > \quad \angle ABC + \angle ACB$$

By [1.13], $\angle ACD + \angle ACB$ equals two right angles; therefore, $\angle ABC + \angle ACB$ is less than two right angles.

Similarly, we can show that the sums $\angle ABC + \angle BAC$ and $\angle ACB + \angle BAC$ are each less than two right angles, *mutatis mutandis*. A similar argument follows on $\triangle EFG$, which proves our claim. □

Corollary. *1.17.1. Every triangle has at least two acute angles.*

Corollary. *1.17.2. If two angles of a triangle are unequal, then the shorter angle is acute.*

Exercises.

1. Prove [1.17] without extending a side. (Attempt after completing Chapter 1. Hint: use parallel line theorems.)

2. Prove [Cor. 1.17.1].

3. Prove [Cor. 1.17.2].

Proposition 1.18. *ANGLES AND SIDES IN A TRIANGLE I.*

If one side of a triangle is longer than another side, then the angle opposite the longer side is greater in measure than the angle opposite the shorter side.

Proof. Construct $\triangle ABC$ with sides AB and AC where $AC > AB$. We claim that the angle opposite AC is greater in measure than the angle opposite AB; or, $\angle ABC > \angle ACB$.

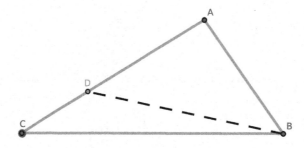

Figure 1.4.20: [1.18]

On AC, find D such that $\overline{AD} = \overline{AB}$ [1.3], and construct $\overline{BD}$; notice that $\triangle ABD$ is isosceles. By [1.6], $\angle ADB = \angle ABD$. Since $\angle ADB > \angle ACB$ by [1.16], $\angle ABD > \angle ACB$. Since $\angle ABC = \angle ABD + \angle CBD$, we also have that $\angle ABC > \angle ACB$, which proves our claim. $\square$

Exercises.

1. Prove that if two of the opposite sides of a quadrilateral are respectively the greatest and the least sides of the quadrilateral, then the angles adjacent to the least are greater than their opposite angles.

2. In any triangle, prove that the perpendicular from the vertex opposite the side which is not less than either of the remaining sides falls within the triangle.

1.4. BOOK I, PROPOSITIONS 1-26

Proposition 1.19. *ANGLES AND SIDES IN A TRIANGLE II.*

In a triangle, if one angle is greater in measure than another, then the side opposite the greater angle is longer than the side opposite the shorter angle.

Proof. Construct $\triangle ABC$ where $\angle ABC > \angle ACB$. We claim that $AC > AB$.

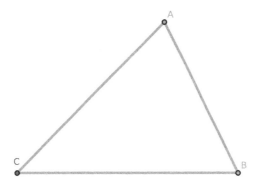

Figure 1.4.21: [1.19]

If $AC \not> AB$, then either $AC = AB$ or $AC < AB$.

1. If $AC = AB$, $\triangle ACB$ is isosceles and $\angle ACB = \angle ABC$ [Cor. 1.6.1]. This contradicts our hypothesis that $\angle ABC > \angle ACB$, and so $AC \neq AB$.

2. If $AC < AB$, we find that $\angle ACB > \angle ABC$ [1.18]. This also contradicts our hypothesis that $\angle ABC > \angle ACB$, and so $AC \not< AB$.

Since $AC \not\leq AB$, we must have that $AC > AB$. □

Corollary. 1.19.1. *In a triangle, longer sides stand opposite the greater interior angles and shorter interior angles stand opposite the shorter sides.*

Exercises.

1. Prove this proposition by a direct demonstration.

2. Prove that a segment from the vertex of an isosceles triangle to any point on the base is less than either of the equal sides but greater if the base is extended and the point of intersection falls outside of the triangle.

3. Prove that three equal and distinct segments cannot be constructed from the same point to the same line. [See the final chapter for a solution.]

4. Consider [1.16], Fig 1.4.18: if AB is the longest side of the $\triangle ABC$, then BF is the longest side of $\triangle FBC$ and $\angle BFC < \frac{1}{2} \cdot \angle ABC$.

5. If $\triangle ABC$ is a triangle such that side $AB \leq AC$, then a segment $\overline{AG}$, constructed from A to any point G on side BC, is less than AC. [See the final chapter for a solution.]

6. Prove [Cor. 1.19.1].

Proposition 1.20. *THE SUM OF THE LENGTHS OF ANY PAIR OF SIDES OF A TRIANGLE.*

In a triangle, the sum of the lengths of any pair of sides is greater than the length of the remaining side.

Proof. Construct $\triangle ABC$; wlog, we claim that $AB + AC > BC$.

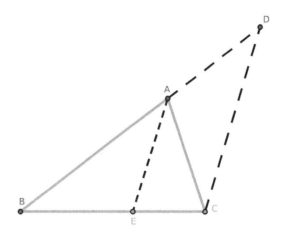

Figure 1.4.22: [3.20]

Extend BA to $\overline{BD}$ such that $\overline{AD} = \overline{AC}$ [1.3], and construct $\overline{CD}$.

Consider $\triangle ACD$: by construction $\overline{AD} = \overline{AC}$, and so $\angle ACD = \angle ADC$ [1.5]. Since $\angle BCD = \angle BCA + \angle ACD$, $\angle BCD > \angle ACD = \angle ADC = \angle BDC$. By [1.19], $\overline{BD} > \overline{BC}$.

Notice that

$$\begin{aligned} \overline{AD} &= \overline{AC} \\ \overline{BA} + \overline{AD} &= \overline{BA} + \overline{AC} \\ \overline{BD} &= \overline{BA} + \overline{AC} \end{aligned}$$

and so $\overline{BA} + \overline{AC} > \overline{BC}$, which proves our claim. $\square$

Alternatively:

Proof. Construct $\triangle ABC$; wlog, we claim that $AB + AC > BC$. Bisect $\angle BAC$ by constructing $\overline{AE}$ [1.9]. Then $\angle BEA > \angle EAC = \angle EAB$. It follows that $\overline{BA} > \overline{BE}$ [1.19]. Similarly, $\overline{AC} > \overline{EC}$. It follows that

$$\overline{BA} + \overline{AC} > \overline{BE} + \overline{EC} = \overline{BC}$$

which proves our claim. $\square$

Exercises.

1. Let a, b, and c be side-lengths of any triangle. Prove that

$$|a - b| < c < (a + b)$$

2. Any side of any polygon is less than the sum of the remaining sides.

3. The perimeter of any triangle is greater than the perimeter of any inscribed triangle and less than the perimeter of any circumscribed triangle. (See also [Def. 4.1].)

4. The perimeter of any polygon is greater than that of any inscribed (and less than that of any circumscribed) polygon of the same number of sides.

5. The perimeter of a quadrilateral is greater than the sum of its diagonals. [See the final chapter for a solution.]

6. The sum of the lengths of the three medians of a triangle is less than $\frac{3}{2}$ times its perimeter. [See the final chapter for a solution.]

1.4. BOOK I, PROPOSITIONS 1-26

Proposition 1.21. *TRIANGLES WITHIN TRIANGLES.*

In an arbitrary triangle, if two segments are constructed from the vertexes of its base to a point within the triangle, then:

(1) the sum of the lengths of these inner sides will be less than the sum of the outer sides (excluding the base);

(2) these inner sides will contain a greater angle than the corresponding sides of the outer triangle.

Proof. Construct $\triangle ABC$ with base BC, and construct D within $\triangle ABC$. Finally, construct segments $\overline{BD}, \overline{CD}$. We claim that:

1. $BA + AC > BD + DC$
2. $\angle BDC > \angle BAC$

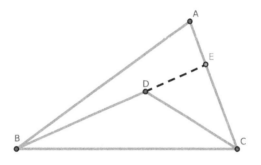

Figure 1.4.23: [3.21]

Claim 1: $BA + AC > BD + DC$.

Extend BD to $\overline{BE}$ where E is on AC. In $\triangle BAE$, we find that $BA + AE > BE$ [1.20]. It follows that:
$$BA + AC = BA + AE + EC > BE + EC$$

Similarly, in $\triangle DEC$, we find that $DE + EC > DC$, from which it follows that
$$BE + EC = BD + DE + EC > BD + DC$$

From these two inequalities, we obtain that $BA + AC > BD + DC$, which proves claim 1.

Claim 2: $\angle BDC > \angle BAC$.

By [1.16], we find that $\angle BDC > \angle BEC$. Similarly, $\angle BEC > \angle BAE$. It follows that $\angle BDC > \angle BAE = \angle BAC$, which proves claim 2 and completes the proof. □

An alternative proof of claim 2 that does not extend sides BD and DC:

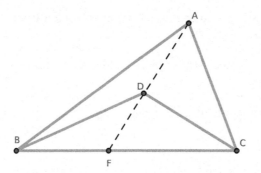

Figure 1.4.24: [3.21, alternate proof]

Proof. Construct $\triangle ABC$ and $\triangle BDC$ as above. Also construct $\overline{AD}$ and extending it to intersect BC at F. Consider $\triangle BDA$ and $\triangle CDA$. By [1.16], $\angle BDF > \angle BAF$ and $\angle FDC > \angle FAC$. Then

$$\angle BDC = \angle BDF + \angle FDC > \angle BAF + \angle FAC = \angle BAC$$

which completes the proof. $\square$

Exercises.

1. The sum of the side lengths constructed from any point within a triangle to its vertices is less than the length of the triangle's perimeter.

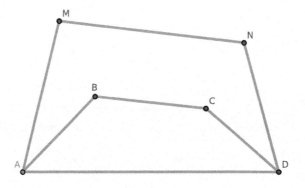

Figure 1.4.25: [1.21, #2]

2. If a convex polygonal line $ABCD$ lies within a convex polygonal line $AMND$ terminating at the same endpoints, prove that the length of $ABCD$ is less than that of $AMND$.

1.4. BOOK I, PROPOSITIONS 1-26

Proposition 1.22. *CONSTRUCTION OF TRIANGLES FROM ARBITRARY SEGMENTS.*

It is possible to construct a triangle whose sides are respectively equal in length to three arbitrary segments whenever the sum of the lengths of each pair of segments is greater than the length of the remaining segment.

Proof. Let $\overline{AR}$, $\overline{BS}$, and $\overline{CT}$ be arbitrary segments which satisfy our hypothesis.

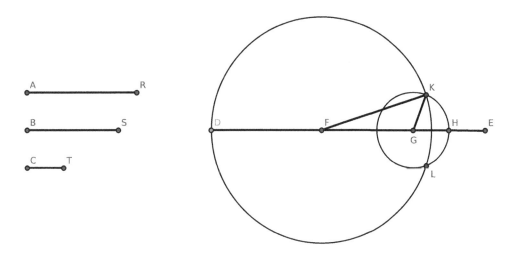

Figure 1.4.26: [1.22]

Construct $\overrightarrow{DE}$ such that it contains the segments $\overline{DF} = \overline{AR}$, $\overline{FG} = \overline{BS}$, and $\overline{GH} = \overline{CT}$ [1.3]. With F as the center and $\overline{DF}$ as radius, construct $\odot F$ [Section 1.2, Postulate 3]. With G as the center and $\overline{GH}$ as radius, construct $\odot G$ where K is one intersection between $\odot F$ and $\odot G$. Construct $\overline{KF}$, $\overline{KG}$. We claim that $\triangle KFG$ is the required triangle.

Since F is the center of $\odot F$, $\overline{FK} = \overline{FD}$. Since $\overline{FD} = \overline{AR}$ by construction, $\overline{FK} = \overline{AR}$. Also by construction, $\overline{FG} = \overline{BS}$ and $\overline{KG} = \overline{CT}$. Therefore, the three sides of the triangle $\triangle KFG$ are respectively equal to the three segments $\overline{AR}$, $\overline{BS}$, and $\overline{CT}$, which proves our claim. □

Exercises.

1. Prove that when the above conditions are fulfilled that the two circles must intersect.

2. If the sum of two of the segments equals the length of the third, prove that the segments will not intersect.

68   CHAPTER 1. ANGLES, PARALLEL LINES, PARALLELOGRAMS

Proposition 1.23. CONSTRUCTING EQUAL ANGLES.

It is possible to construct an angle equal to an arbitrary angle on a given point.

Proof. Construct an arbitrary angle $\angle DEF$ from $\overrightarrow{ED}$ and $\overrightarrow{EF}$ as well as point A. We claim it is possible to construct an angle equal to $\angle DEF$ on A.

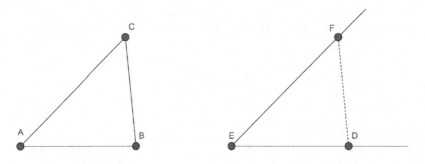

Figure 1.4.27: [1.23]

Construct $\overline{DF}$, and construct the triangle $\triangle BAC$ where $\overline{AB} = \overline{ED}$, $\overline{AC} = \overline{EF}$, and $\overline{CB} = \overline{FD}$ [1.22]. By [1.8], $\triangle BAC \cong \triangle DEF$, and so $\angle BAC = \angle DEF$. □

Exercises.

1. Construct a triangle given two sides and the angle between them. [See the final chapter for a solution.]

2. Construct a triangle given two angles and the side between them.

3. Construct a triangle given two sides and the angle opposite one of them.

4. Construct a triangle given the base, one of the angles at the base, and the sum or difference of the sides.

5. Given two points, one of which is in a given line, find another point on the given line such that the sum or difference of its distances from the former points may be given. Show that two such points may be found in each case.

1.4. BOOK I, PROPOSITIONS 1-26

Proposition 1.24. *ANGLES AND SIDES IN A TRIANGLE III.*

Suppose we have two triangles such that two sides of the first triangle are respectively equal in length to two sides of the second triangle and that the interior angles of each of these pairs of sides are unequal. The third side of the triangle with the larger interior angle will be longer than the third side of the triangle with the smaller interior angle.

Proof. Construct two triangles $\triangle ABC$ and $\triangle DEF$ where $AB = DE$, $AC = DF$, and $\angle BAC > \angle EDF$. We claim that $BC > EF$.

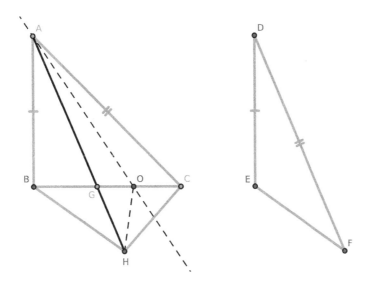

Figure 1.4.28: [1.24]

Construct point G on BC such that $\angle BAG = \angle EDF$. Wlog, suppose that $AB < AC$; by [1.19, #5] we find that $\overline{AG} < \overline{AC}$. Extend $\overline{AG}$ to $\overline{AH}$ where $\overline{AH} = \overline{DF} = \overline{AC}$ [1.3].

Construct $\overline{BH}$ and $\overline{CH}$. In triangles $\triangle BAH$ and $\triangle EDF$, we have $AB = DE$, $AH = DF$, and $\angle BAH = \angle EDF$ by construction. By [1.4], $\triangle BAH \cong \triangle EDF$, and so $BH = EF$.

Notice that $\angle ACH > \angle BCH$ since $\angle ACH = \angle BCH + \angle BCA$. Because $AH = AC$ by construction, $\triangle ACH$ is isosceles; therefore, $\angle ACH = \angle AHC$ [1.5]. It follows that $\angle AHC > \angle BCH$. And since $\angle BHC = \angle BHA + \angle AHC$, we also have that $\angle BHC > \angle BCH$.

By [1.19], $BC > BH$. Since $BH = EF$ by the above, $BC > EF$, which proves our claim. $\square$

Alternatively, the concluding part of this proposition may be proved without constructing $\overline{CH}$.

Proof. Assume the hypotheses and construct the triangles as in the previous proof; we claim that this proof does not require $\overline{CH}$.

Notice that

$$\begin{aligned} BG + GH &> BH & [1.20] \quad &\text{and} \\ AG + GC &> AC & [1.20] \quad &\Rightarrow \\ (BG + GH) + (AG + GC) &> BH + AC & &\Rightarrow \\ (BG + GC) + (AG + GH) &> BH + AC & &\Rightarrow \\ BC + AH &> BH + AC & & \end{aligned}$$

Since $AH = AC$ and $BH = EF$ by construction, we have

$$\begin{aligned} BC + AC &> EF + AC \\ BC &> EF \end{aligned}$$

which proves our claim. □

Another alternative:

Proof. In $\triangle ABC$, bisect the angle $\angle CAH$ by $\overleftrightarrow{AO}$. In $\triangle CAO$ and $\triangle HAO$ we have the sides CA, AO in one triangle respectively equal to the sides AH, AO in the other where the interior angles are equal. By [1.4], $OC = OH$. It follows that $BO + OH = BO + OC = BC$. But $BO + OH > BH$ [1.20], and so $BC > BH$. Since $BH = EF$, $BC > EF$, which proves our claim. □

Remark. The reader will have noticed by now that the number of explicit references to definitions, axioms, and theorems has dwindled since the first proposition. This is normal in mathematical writing even though it is sub-optimal to the reader. Such practice is normal because to do otherwise would place an impossible burden on the writer. One reader will wish I had cited the Inequality Properties of Addition and Subtraction in the above. Another would be insulted if I had halted the proof to cite something so obvious.[36] Yet another reader will ask why the Inequality Properties of Addition and Subtraction was assumed rather than proven.[37] No writer can satisfy contradictory demands. The best way out for the author is to discard whatever he or she finds obvious and to leave the detective work to the reader.

---

[36] Exactly what is and is not obvious in mathematics is a can of worms I do not intend to open here.

[37] The questions regarding the foundations of mathematics are not as simple as one might think. Whitehead & Russell's "Principia Mathematica" is so detailed that the authors take well over 300 pages to prove that $1 + 1 = 2$.

Exercises.

1. Prove this proposition by constructing the angle $\angle ABH$ to the left of AB.

2. Prove that the angle $\angle BCA > \angle EFD$.

Proposition 1.25. *ANGLES AND SIDES IN A TRIANGLE IV.*

Suppose two triangles exist such that two sides of the first triangle are respectively equal in length to two sides of the second triangle and the third sides are unequal in length. The triangle with the longer third side will have a larger interior angle than the triangle with the shorter side.

Proof. Construct $\triangle ABC$ and $\triangle DEF$ such that $AB = DE$, $AC = DF$, and $BC > EF$. We claim that $\angle BAC > \angle EDF$.

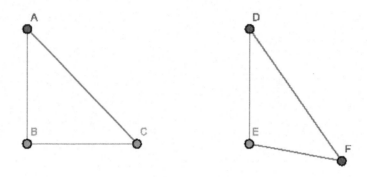

Figure 1.4.29: [1.25]

Suppose instead that $\angle BAC = \angle EDF$; since $AB = DE$ and $AC = DF$ by construction, by [1.4] $\triangle ABC \cong \triangle DEF$, and so $BC = EF$. This contradicts our hypothesis that $BC > EF$; hence, $\angle BAC \neq \angle EDF$.

Now suppose that $\angle BAC < \angle EDF$; since $AB = DE$ and $AC = DF$, by [1.24] we find that $EF > BC$. This contradicts our hypothesis that $BC > EF$; hence, $\angle BAC \not< \angle EDF$.

Since $\angle BAC \not< \angle EDF$, $\angle BAC > \angle EDF$, which proves our claim. □

Corollary. *1.25.1.* Construct $\triangle ABC$ and $\triangle DEF$ where $AB = DE$ and $AC = DF$. By [1.24] and [1.25], $BC > EF$ iff $\angle BAC > \angle EDF$.

Exercise.

1. Demonstrate this proposition directly by constructing a segment on BC equal in length to EF.

1.4. BOOK I, PROPOSITIONS 1-26

Proposition 1.26. *CONGRUENCE OF TRIANGLES WITH ONE EQUAL SIDE AND TWO EQUAL INTERIOR ANGLES.*

If two triangles exist such that one side of the first triangle is equal in length to one side of the second triangle and that two interior angles of the first triangle are respectively equal in measure to two interior angles of the second triangle, then the triangles are congruent. Most express this proposition as two theorems:

1. "ANGLE-SIDE-ANGLE" CONGRUENCE. If the equal side stands between the two equal angles, then the triangles are congruent.

2. "ANGLE-ANGLE-SIDE" CONGRUENCE. If the equal side does not stand between the two equal angles, then the triangles are congruent.

We will prove this proposition in two cases.

Proof. Construct $\triangle ABC$ and $\triangle DEF$ such that $\angle ABC = \angle DEF$ and $\angle ACB = \angle EFD$. We claim that if one side of $\triangle ABC$ is equal in length to its respective side in $\triangle DEF$, then $\triangle ABC \cong \triangle DEF$.

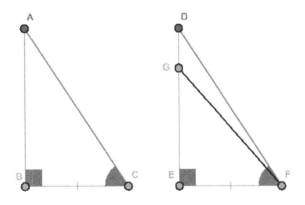

Figure 1.4.30: [1.26], case 1

Case 1. *ANGLE-SIDE-ANGLE*

Suppose that $BC = EF$. If $AB \neq DE$, suppose that $AB = GE$ where G is a point on $\overline{DE}$ such that $D \neq G$. Construct $\overline{GF}$, and notice that $\angle GFD > 0$. (If $\angle GFD = 0$, then $D = G$, a contradiction.)

Consider $\triangle ABC$ and $\triangle GEF$: $AB = GE$, $BC = EF$, and $\angle ABC = \angle GEF$. By [1.4], $\triangle ABC \cong \triangle GEF$, and so $\angle ACB = \angle GFE$. Since $\angle ACB = \angle DFE$ by hypothesis, we find that $\angle GFE = \angle DFE$ and $\angle GFE + \angle GFD = \angle DFE$; hence $\angle GFD = 0$ and $\angle GFD > 0$, a contradiction. Therefore, $AB = DE$.

Since $AB = DE$, $BC = EF$, and $\angle ABC = \angle DEF$, by [1.4] $\triangle ABC \cong \triangle DEF$.

Case 2. *ANGLE-ANGLE-SIDE*

Now suppose that $AB = DE$.

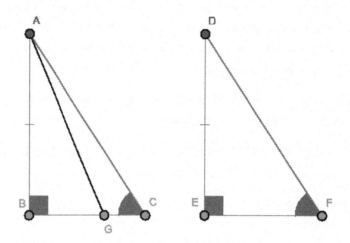

Figure 1.4.31: [1.26], case 2

If $BC \neq EF$, suppose that $EF = BG$ where G is a point on $\overline{BC}$ such that $C \neq G$. Construct $\overline{AG}$, and consider $\triangle ABG$ and $\triangle DEF$: $AB = DE$, $BG = EF$, and $\angle ABG = \angle DEF$. By [1.4], $\triangle ABG \cong \triangle DEF$, and so $\angle AGB = \angle DFE$. Since $\angle ACB = \angle DFE$ by hypothesis, $\angle AGB = \angle ACB$; that is, the exterior angle of $\triangle ACG$ is equal to an interior and non-adjacent angle, contradicting [1.16]. Thus $BC = EF$.

Since $AB = DE$, $BC = EF$, and $\angle ABC = \angle DEF$, by [1.4] $\triangle ABC \cong \triangle DEF$.

This proves both claims and completes proof. □

Exercises.

1. Prove that the endpoints of the base of an isosceles triangle are equally distant from any point on the perpendicular segment from the vertical angle on the base.

2. Prove that if the line which bisects the vertical angle of a triangle also bisects the base, then the triangle is isosceles.

3. In a given straight line, find a point such that the perpendiculars from it on two given lines are equal. State also the number of solutions.

4. Prove that if two right triangles have hypotenuses of equal length and an acute angle of one is equal to an acute angle of the other, then they are congruent.

5. Prove that if two right triangles have equal hypotenuses and that if a side of one is equal in length to a side of the other, then the triangles they are congruent. (Note: this proves the special case of Side-Side-Angle congruency for right triangles.)

6. The bisectors of two external angles and the bisector of the third internal angle are concurrent.

7. Through a given point, construct a straight line such that perpendiculars on it from two given points on opposite sides are equal to each other.

8. Through a given point, construct a straight line intersecting two given lines which forms an isosceles triangle with them.

1.5 Book I, Propositions 27-48

Additional definitions regarding parallel lines:

Parallel Lines

40. If two straight lines in the same plane do not meet at any finite distance, they are said to be *parallel*. If rays or segments can be extended into lines which do not meet at any finite distance, they are also said to be *parallel*.

41. A *parallelogram* is a quadrilateral where both pairs of opposite sides are parallel.

42. The segment joining either pair of opposite angles of a quadrilateral is called a *diagonal*. See Fig. 1.5.1.

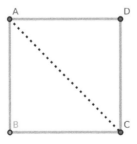

Figure 1.5.1: [Def 1.41] and [Def. 1.42]: $\overline{AC}$ is a diagonal of the square $\square ABCD$, which is also a parallelogram.

43. The *altitude* of a triangle is the perpendicular segment from the triangle's base to the base's opposing vertex.

45. A quadrilateral where one pair of opposite sides is parallel is called a *trapezoid*.

76 CHAPTER 1. ANGLES, PARALLEL LINES, PARALLELOGRAMS

46. When a straight line intersects two other straight lines, between them are eight angles (see Fig. 1.5.2).

- Angles 1 and 2 are *exterior angles*; so are angles 7 and 8.

- Angles 3 and 4 are called *interior angles*; so are angles 5 and 6.

- Angles 4 and 6 are called *alternate angles;* so are angles 3 and 5.

- Angles 1 and 5 are called *corresponding angles*; so are angles 2 and 6, 3 and 8, and 4 and 7.

These definitions hold if we replace lines with either rays or segments, *mutatis mutandis*.

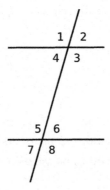

Figure 1.5.2: [Def. 1.46]

1.5. BOOK I, PROPOSITIONS 27-48

Proposition 1.27. *PARALLEL LINES I.*

If a line intersects a pair of lines and their alternate angles are equal, then the pair of lines are parallel.

Proof. Let $\overleftrightarrow{EF}$ intersect $\overleftrightarrow{AB}$ and $\overleftrightarrow{CD}$ such that $\angle AEF = \angle EFD$. We claim that $\overleftrightarrow{AB} \parallel \overleftrightarrow{CD}$.

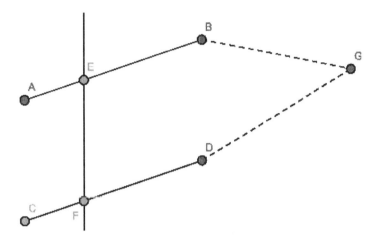

Figure 1.5.3: [1.27]

If $\overleftrightarrow{AB} \not\parallel \overleftrightarrow{CD}$, then $\overleftrightarrow{AB}$ and $\overleftrightarrow{CD}$ intersect at point G where the length of $\overline{BG}$ is finite.[38]

It follows that $\triangle EGF$ is a triangle where $\angle AEF$ is an exterior angle and $\angle EFG$ a non-adjacent interior angle. By [1.16], $\angle AEF > \angle EFD$; but $\angle AEF = \angle EFD$ by hypothesis, a contradiction. Therefore, $\overleftrightarrow{AB} \parallel \overleftrightarrow{CD}$. □

---

[38]"The length of $\overline{BG}$ is finite" can also be expressed as "$\overline{BG} < \infty$".

CHAPTER 1. ANGLES, PARALLEL LINES, PARALLELOGRAMS

Proposition 1.28. *PARALLEL LINES II & III.*

Suppose a line intersects a pair of lines.

1) PARALLEL LINES II. If the intersecting line makes the exterior angle equal to its corresponding interior angle, then the pair of lines is parallel.

2) PARALLEL LINES III. If the intersecting line makes the sum of two interior angles on the same side equal to two right angles, then the pair of lines is parallel.

Proof. Suppose $\overleftrightarrow{EF}$ intersects $\overleftrightarrow{AB}$ and $\overleftrightarrow{CD}$.

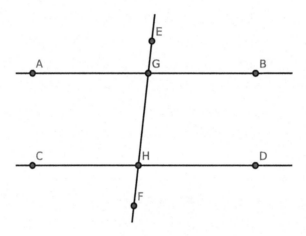

Figure 1.5.4: [1.28] and [1.29]

Claim 1: If $\angle EGB = \angle GHD$, we claim that $\overleftrightarrow{AB} \parallel \overleftrightarrow{CD}$.

Since $\overleftrightarrow{AB}$, $\overleftrightarrow{EF}$ intersect at G, $\angle AGH = \angle EGB$ [1.15]. By hypothesis, $\angle AGH = \angle GHD$. Since these are alternate angles, $\overleftrightarrow{AB} \parallel \overleftrightarrow{CD}$ by [1.27], which proves our claim.

Claim 2: If $\angle BGH + \angle GHD$ equals two right angles, we claim that $\overleftrightarrow{AB} \parallel \overleftrightarrow{CD}$.

Since $\angle AGH$ and $\angle BGH$ are adjacent angles, by [1.13] the sum $\angle AGH + \angle BGH$ equals two right angles. Since $\angle BGH = \angle BGH$,

$$\angle BGH + \angle GHD = \angle BGH + \angle AGH$$
$$\angle GHD = \angle AGH$$

Since $\angle GHD$ and $\angle AGH$ are alternate angles, $\overleftrightarrow{AB} \parallel \overleftrightarrow{CD}$ by [1.27], which proves our claim and completes the proof. □

Proposition 1.29. *PARALLEL LINES IV.*

If a line intersects two parallel lines, then:

(1) corresponding alternate angles are equal,

(2) exterior angles are equal to corresponding interior angles,

(3) the sum of interior angles on the same side equals two right angles.

Proof. If $\overleftrightarrow{EF}$ intersects $\overleftrightarrow{AB}$ and $\overleftrightarrow{CD}$ where $\overleftrightarrow{AB} \parallel \overleftrightarrow{CD}$, we claim that:

(1) $\angle AGH = \angle GHD$ ($\angle BGH = \angle GHC$, *mutatis mutandis*);

(2) $\angle EGB = \angle GHD$ ($\angle EGA = \angle GHC$, *mutatis mutandis*);

(3) $\angle GHD + \angle HGB$ equals two right angles ($\angle AGH + \angle GHC$ also equals two right angles, *mutatis mutandis*)

Claim 1: if $\angle AGH \neq \angle GHD$, one angle must be greater than the other. Wlog, suppose that $\angle AGH > \angle GHD$. Then we obtain the inequality

$$\angle AGH + \angle BGH > \angle GHD + \angle BGH$$

where $\angle AGH + \angle BGH$ is equal to the sum of two right angles by [1.13]. It follows that $\angle GHD + \angle BGH$ is less than two right angles. By the proof of [1.27], $\overleftrightarrow{AB}$ and $\overleftrightarrow{CD}$ meet at some finite distance; this contradicts our hypothesis that $\overleftrightarrow{AB} \parallel \overleftrightarrow{CD}$. Hence, $\angle AGH = \angle GHD$, proving claim 1.

Claim 2: since $\angle EGB = \angle AGH$ by [1.15] and $\angle AGH = \angle GHD$ by claim 1, $\angle EGB = \angle GHD$, proving claim 2.

Claim 3. since $\angle AGH = \angle GHD$ by claim 1,

$$\angle AGH + \angle HGB = \angle GHD + \angle HGB$$

Since $\angle AGH + \angle HGB$ equals the sum of two right angles, $\angle GHD + \angle HGB$ equals the sum of two right angles. This proves claim 3 and completes the proof. □

80 CHAPTER 1. ANGLES, PARALLEL LINES, PARALLELOGRAMS

Corollary. *1.29.1. EQUIVALENT STATEMENTS REGARDING PARALLEL LINES. If a line intersects a pair of lines, then the pair of lines are parallel if and only if any of these three properties hold:*

1) corresponding alternate angles are equal;

2) exterior angles equal their corresponding interior angles;

3) the sum of the interior angles on the same side are equal to two right angles.

Corollary. *1.29.2. We may replace the lines in [1.29, Cor. 1] with segments of appropriate length or rays, mutatis mutandis.*

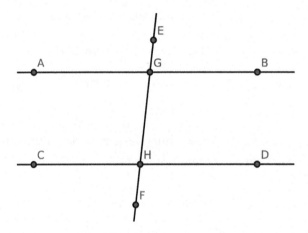

Figure 1.5.5: [1.28] and [1.29]

Exercises.

Remark. We may use [1.31] in the proofs of these exercises since the proof of [1.31] does not require [1.29].

1. Demonstrate both parts of [1.28] without using [1.27].

2. Construct $\overleftrightarrow{AB}$ containing the point C and $\overleftrightarrow{EF}$ containing the point D such that $\overleftrightarrow{AB} \parallel \overleftrightarrow{EF}$. Construct $\overline{CH}$ and $\overline{CJ}$ such that $\overline{CJ}$ bisects $\angle ACD$ and $\overline{CH}$ bisects $\angle BCD$. Prove that $\overline{DH} = \overline{DJ}$. [See the final chapter for a solution.]

4. If any other secant is constructed through the midpoint O of any line terminated by two parallel lines, the intercept on this line made by the parallels is bisected at O.

5. Two lines passing through a point which is equidistant from two parallel lines intercept equal segments on the parallels. [See the final chapter for a solution.]

1.5. BOOK I, PROPOSITIONS 27-48

6. Construct the perimeter of the parallelogram formed by constructing parallels to two sides of an equilateral triangle from any point in the third side. This perimeter is equal to 2× the side.

7. If the opposite sides of a hexagon are equal and parallel, prove that its diagonals are concurrent.

8. If two intersecting segments are respectively parallel to two others, the angle between the former is equal to the angle between the latter. (Hint: if $\overleftrightarrow{AB}$, $\overleftrightarrow{AC}$ are respectively parallel to $\overleftrightarrow{DE}$, $\overleftrightarrow{DF}$ and if $\overleftrightarrow{AC}$, $\overleftrightarrow{DE}$ intersect at G, then the angles at points A, D are each equal to the angle at G [1.29].)

Proposition 1.30. *TRANSITIVITY OF PARALLEL LINES.*

Lines parallel to the same line are parallel to each other.

Proof. Construct lines $\overleftrightarrow{AB}$, $\overleftrightarrow{CD}$, $\overleftrightarrow{EF}$ such that $\overleftrightarrow{AB} \parallel \overleftrightarrow{EF}$ and $\overleftrightarrow{CD} \parallel \overleftrightarrow{EF}$. We claim that $\overleftrightarrow{AB} \parallel \overleftrightarrow{CD}$.

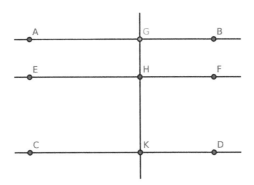

Figure 1.5.6: [1.30]

Construct any secant line $\overleftrightarrow{GHK}$. Since $\overleftrightarrow{AB} \parallel \overleftrightarrow{EF}$, the angle $\angle AGH = \angle GHF$ [Cor. 1.29.1]. Since $\overleftrightarrow{CD} \parallel \overleftrightarrow{EF}$, the angle $\angle GHF = \angle HKD$ [Cor. 1.29.1]. It follows that $\angle AGH = \angle GHF = \angle HKD$. Since $\angle AGH = \angle AGK$ and $\angle HKD = \angle GKD$, $\angle AGK = \angle GKD$. By [1.27], $\overleftrightarrow{AB} \parallel \overleftrightarrow{CD}$. □

Corollary. *1.30.1.* $\overleftrightarrow{AB}$, $\overleftrightarrow{CD}$, $\overleftrightarrow{EF}$, and $\overleftrightarrow{GK}$ in [1.30] may be replaced by segments of appropriate length or rays, mutatis mutandis.

CHAPTER 1. ANGLES, PARALLEL LINES, PARALLELOGRAMS

Proposition 1.31. *CONSTRUCTION OF A PARALLEL LINE.*

We wish to construct a line which is parallel to a given line and passing through a given point.

Proof. Given the line $\overleftrightarrow{AB}$ and a point C, we wish to construct the line $\overleftrightarrow{CE}$ such that $\overleftrightarrow{CE} \parallel \overleftrightarrow{AB}$.

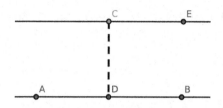

Figure 1.5.7: [1.31]

Using [1.23], construct D on $\overleftrightarrow{AB}$ and E not on $\overleftrightarrow{AB}$ such that, after constructing $\overleftrightarrow{CE}$, $\angle ADC = \angle DCE$. By [Cor. 1.29.1], $\overleftrightarrow{CE} \parallel \overleftrightarrow{AB}$. □

Corollary. *1.31.1.* $\overleftrightarrow{AB}$ *and* $\overleftrightarrow{CE}$ *in [1.31] may be replaced by segments of appropriate length or rays, mutatis mutandis.*

Exercises.

1. Given the altitude of a triangle and the base angles, construct the triangle. [See the final chapter for a solution.]

2. From a given point, construct a segment to a given segment such that the resultant angle is equal in measure to a given angle. Show that there are two solutions.

3. Prove the following construction for trisecting a given line $\overleftrightarrow{AB}$: on $\overleftrightarrow{AB}$, construct an equilateral $\triangle ABC$. Bisect the angles at points A and B by the lines $\overleftrightarrow{AD}$ and $\overleftrightarrow{BD}$. Through D, construct parallels to $\overleftrightarrow{AC}$ and $\overleftrightarrow{BC}$ which intersect $\overleftrightarrow{AB}$ at E and F. Claim: E and F are the points of trisection of $\overleftrightarrow{AB}$.

4. Inscribe a square in a given equilateral triangle such that its base stands on a given side of the triangle.

5. Through two given points on two parallel lines, construct two segments forming a rhombus with given parallels.

6. Between two lines given in position, place a segment of given length which is parallel to a given line. Show that there are two solutions.

Proposition 1.32. *EXTERIOR ANGLES AND SUMS OF ANGLES IN A TRIANGLE.*

If the side of a triangle is extended,

(1) the exterior angle equals the sum of the its interior and opposite angles;

(2) the sum of the three interior angles equals two right angles.

Proof. Construct $\triangle ABC$, and wlog extend side AB to segment $\overline{AD}$.

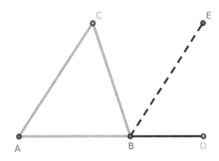

Figure 1.5.8: [1.32]

Claim 1: $\angle CBD = \angle BAC + \angle ACB$.

Construct $\overline{BE} \parallel \overline{AC}$ [1.31]. Since BC intersects $\overline{BE}$ and $\overline{AC}$, we find that $\angle EBC = \angle ACB$ [1.29]. Also, since AB intersects $\overline{BE}$ and $\overline{AC}$, we find that $\angle DBE = \angle BAC$ [1.29]. Since $\angle CBD = \angle EBC + \angle DBE$,

$$\angle CBD = \angle ACB + \angle BAC$$

Claim 2: $\angle BAC + \angle ACB + \angle ABC =$ two right angles.

Adding $\angle ABC$ to each side of the equality in claim 1, we obtain

$$\angle ABC + \angle CBD = \angle ABC + \angle ACB + \angle BAC$$

By [1.13], $\angle ABC + \angle CBD$ equals two right angles, and so

$$\angle BAC + \angle ACB + \angle ABC$$

equals two right angles. This completes the proof. □

Corollary. *1.32.1. If a right triangle is isosceles, then each base angle equals half of a right angle.*

Corollary. *1.32.2. If two triangles have two angles in one respectively equal to two angles in the other, then their remaining pair of angles is also equal.*

Corollary. *1.32.3. Since a quadrilateral can be divided into two triangles, the sum of its angles equals four right angles.*

Corollary. *1.32.4. If a figure of n sides is divided into triangles by drawing diagonals from any one of its angles, we will obtain $(n-2)$ triangles. Hence, the sum of its angles equals $2(n-2)$ right angles.*

Corollary. *1.32.5. If all the sides of any convex polygon are extended, then the sum of the external angles equals to four right angles.*

Corollary. *1.32.6. Each angle of an equilateral triangle equals two-thirds of a right angle.*

Corollary. *1.32.7. If one angle of a triangle equals the sum of the other two, then it is a right angle.*

Corollary. *1.32.8. Every right triangle can be divided into two isosceles triangles by a line constructed from the right angle to the hypotenuse.*

Exercises.

1. Trisect a right angle.

2. If the sides of a polygon of n sides are extended, then the sum of the angles between each alternate pair is equal to $2(n-4)$ right angles.

3. If the line which bisects an external vertical angle of a triangle is parallel to the base of the triangle, then the triangle is isosceles. [See the final chapter for a solution.]

4. If two right triangles $\triangle ABC$, $\triangle ABD$ are on the same hypotenuse AB and if the vertices C and D are joined, then the pair of angles standing opposite any side of the resulting quadrilateral are equal.

5. Prove that the three altitudes of a triangle are concurrent. Note: We are proving the existence of the **orthocenter**[39] of a triangle: the point where the three altitudes

---

[39]http://mathworld.wolfram.com/Orthocenter.html

1.5. BOOK I, PROPOSITIONS 27-48

intersect, and one of a triangle's **points of concurrency**[40]. [See the final chapter for a solution.]

6. The bisectors of the adjacent angles of a parallelogram stand at right angles. [See the final chapter for a solution.]

7. The bisectors of the external angles of a quadrilateral form a circumscribed quadrilateral, the sum of whose opposite angles equals two right angles.

8. If the three sides of one triangle are respectively perpendicular to those of another triangle, the triangles are equiangular. (This problem may be delayed until the end of chapter 1.)

9. Construct a right triangle being given the hypotenuse and the sum or difference of the sides.

10. The angles made with the base of an isosceles triangle by altitudes from its endpoints on the equal sides are each equal to half the vertical angle.

11. The angle included between the internal bisector of one base angle of a triangle and the external bisector of the other base angle is equal to half the vertical angle.

12. In the construction of [1.18], prove that the angle $\angle DBC$ is equal to half the difference of the base angles.

13. If A, B, C denote the angles of a triangle, prove that $\frac{1}{2}(A+B)$, $\frac{1}{2}(B+C)$, and $\frac{1}{2}(A+C)$ are the angles of a triangle formed by any side, the bisectors of the external angles between that side, and the other extended sides.

14. Prove [Cor. 1.32.1].

15. Prove [Cor. 1.32.2].

16. Prove [Cor. 1.32.3].

17. Prove [Cor. 1.32.4].

18. Prove [Cor. 1.32.5].

19. Prove [Cor. 1.32.6].

20. Prove [Cor. 1.32.7].

21. Prove [Cor. 1.32.8].

---

[40] http://www.mathopenref.com/concurrentpoints.html

Proposition 1.33. *PARALLEL SEGMENTS.*

Segments which join adjacent endpoints of two equal, parallel segments are themselves parallel and equal in length.

Proof. Suppose that $\overline{AB} \parallel \overline{CD}$ and $\overline{AB} = \overline{CD}$. Construct $\square ABDC$. We claim that $\overline{AC} = \overline{BD}$ and $\overline{AC} \parallel \overline{BD}$.

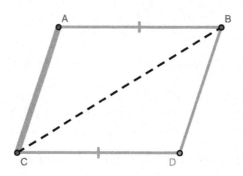

Figure 1.5.9: [1.33] and [1.34]

Construct $\overline{BC}$. Since $\overline{AB} \parallel \overline{CD}$ by hypothesis and $\overline{BC}$ intersects them, $\angle ABC = \angle DCB$ [Cor. 1.29.1].

Consider $\triangle ABC$ and $\triangle DCB$: $AB = DC$, the triangles share side BC, and $\angle ABC = \angle DCB$. By [1.4], $\triangle ABC \cong \triangle DCB$.

It follows that $\overline{AC} = \overline{BD}$ and $\angle ACB = \angle CBD$. By [Cor. 1.29.1], $\angle ACB = \angle CBD$ implies that $\overline{AC} \parallel \overline{BD}$, which completes the proof. $\square$

Corollary. *1.33.1. [1.33] holds for straight lines and rays, mutatis mutandis.*

Corollary. *1.33.2. Figure $\square ABDC$ is a parallelogram [Def. 1.39].*

Exercises.

1. Prove that if two segments $\overline{AB}, \overline{BC}$ are respectively equal and parallel to two other segments $\overline{DE}, \overline{EF}$, then the segment $\overline{AC}$ joining the endpoints of the former pair is equal in length to the segment $\overline{DF}$ joining the endpoints of the latter pair. [See the final chapter for a solution.]

1.5. BOOK I, PROPOSITIONS 27-48

Proposition 1.34. *OPPOSITE SIDES AND OPPOSITE ANGLES OF PARALLELOGRAMS.*

Opposite sides and opposite angles of a parallelogram are equal to one another and either diagonal bisects the parallelogram.

Proof. Construct $\square ABCD$. We claim that:

(1) $\overline{AB} = \overline{CD}$ and $\overline{AC} = \overline{BD}$;

(2) $\angle CAB = \angle CDB$;

(3) $\angle ACD = \angle ABD$;

(4) either diagonal ($\overline{BC}$ or $\overline{AD}$) bisects the parallelogram.

Construct $\overline{BC}$, and consider $\triangle ABC$ and $\triangle DBC$: since $\overline{AB} \parallel \overline{CD}$ by construction and $\overline{BC}$ intersects them, $\angle ABC = \angle DCB$ and $\angle ACB = \angle CBD$ [1.29]. Also, $\triangle ABC$ and $\triangle DBC$ share side $\overline{BC}$. By [1.26], $\triangle ABC \cong \triangle DCB$: it follows that $\overline{AB} = \overline{CD}$ and $\overline{AC} = \overline{BD}$ (claim 1) and $\angle CAB = \angle CDB$ (claim 2).

Now $\angle ACD = \angle ACB + \angle DCB$ and $\angle ABD = \angle CBD + \angle ABC$. Since $\angle ACB = \angle CBD$ and $\angle DCB = \angle ABC$ by [Cor. 1.29.1], we obtain

$$\begin{aligned} \angle ACD &= \angle ACB + \angle DCB \\ &= \angle CBD + \angle ABC \\ &= \angle ABD \end{aligned}$$

(claim 3).

Since $\square ABCD = \triangle ABC \oplus \triangle DEF$ and $\triangle ABC \cong \triangle DCB$, $\triangle ABC = \triangle DBC$. It follows that $\overline{BC}$ bisects $\square ABCD$. The remaining case follows *mutatis mutandis* if we construct $\overline{AD}$ instead of $\overline{BC}$ (claim 4). $\square$

Corollary. *1.34.1.* $\square ABDC = 2 \cdot \triangle ACB = 2 \cdot \triangle BCD$

Corollary. *1.34.2. If one angle of a parallelogram is a right angle, each of its angles are right angles.*

Corollary. *1.34.3. If two adjacent sides of a parallelogram are equal in length, then it is a rhombus.*

Corollary. *1.34.4. If both pairs of opposite sides of a quadrilateral are equal in length, it is a parallelogram.*

Corollary. *1.34.5. If both pairs of opposite angles of a quadrilateral are equal, it is a parallelogram.*

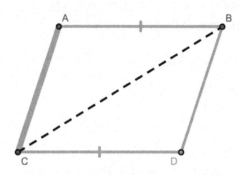

Corollary. *1.34.6. If the diagonals of a quadrilateral bisect each other, it is a parallelogram.*

Corollary. *1.34.7. If both diagonals of a quadrilateral bisect the quadrilateral, it is a parallelogram.*

Corollary. *1.34.8. If the adjacent sides of a parallelogram are equal, its diagonals bisect its angles.*

Corollary. *1.34.9. If the adjacent sides of a parallelogram are equal, its diagonals intersect at right angles.*

Corollary. *1.34.10. In a right parallelogram, the diagonals are equal in length.*

Corollary. *1.34.11. If the diagonals of a parallelogram are perpendicular to each other, the parallelogram is a rhombus.*

Corollary. *1.34.12. If a diagonal of a parallelogram bisects the angles whose vertices it joins, the parallelogram is a rhombus.*

Exercises.

1. Prove that the diagonals of a parallelogram bisect each other. [See the final chapter for a solution.]

2. If the diagonals of a parallelogram are equal, then each of its angles are right angles. [See the final chapter for a solution.]

3. The segments joining the adjacent endpoints of two unequal parallel segments will meet when extended on the side of the shorter parallel.

4. If two opposite sides of a quadrilateral are parallel but unequal in length and the other pair are equal but not parallel, then its opposite angles are supplemental.

5. Construct a triangle after being given the midpoints of its three sides.

6. Prove [Cor. 1.34.1].

7. Prove [Cor. 1.34.2].

8. Prove [Cor. 1.34.3].

9. Prove [Cor. 1.34.4].

10. Prove [Cor. 1.34.5].

11. Prove [Cor. 1.34.6].

12. Prove [Cor. 1.34.7].

13. Prove [Cor. 1.34.8].

14. Prove [Cor. 1.34.9].

15. Prove [Cor. 1.34.10].

16. Prove [Cor. 1.34.11].

17. Prove [Cor. 1.34.12].

90   CHAPTER 1. ANGLES, PARALLEL LINES, PARALLELOGRAMS

Proposition 1.35. *AREAS OF PARALLELOGRAMS I.*

Parallelograms on the same base and between the same parallels are equal in area.

Proof. We shall prove three cases.

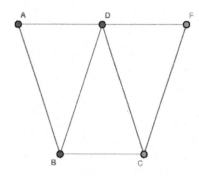

Figure 1.5.10: [1.35], case 1

Case 1: Construct $\square ABCD$ and $\square FDBC$ on base $\overline{BC}$ and between parallels $\overline{AF}$ and $\overline{BC}$ where $\square ABCD$ and $\square FDBC$ share the base and a vertex at D. We claim that $\square ABCD = \square FDBC$.

By [Cor. 1.34.1], $\square ADCB = 2 \cdot \triangle BCD = \square FDBC$, which proves case 1.

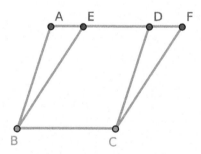

Figure 1.5.11: [1.35], case 2

Case 2: Construct $\square ABCD$ and $\square EFCB$ on base $\overline{BC}$ and between parallels $\overline{AF}$ and $\overline{BC}$ where $\square ABCD$ and $\square EFCB$ share the base and $\overline{ED}$. We claim that $\square ABCD = \square EFCB$.

Because $\square ABCD$ is a parallelogram, $AD = BC$ [1.34]; because $\square BCEF$ is a parallelogram, $EF = BC$, and so $AD = EF$. Notice that

$$\begin{aligned} \overline{AD} - \overline{ED} &= \overline{EF} - \overline{ED} \\ \overline{AE} &= \overline{DF} \end{aligned}$$

Consider $\triangle BAE$ and $\triangle CDF$: $AE = DF$, $BA = CD$ [1.34], and $\angle BAE = \angle CDF$ by [1.29, Cor. 1]. By [1.4], $\triangle BAE \cong \triangle CDF$. Notice that

$$\begin{aligned} AFCB &= \square EFCB + \triangle BAE \\ &= \square ABCD + \triangle CDF \end{aligned}$$

Since $\triangle BAE = \triangle CDF$, $\square ABCD = \square EFCB$, proving case 2.

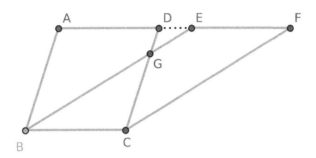

Figure 1.5.12: [1.35], case 3

Case 3: Construct $\square ABCD$ and $\square EFCB$ on base $\overline{BC}$ and between parallels $\overline{AF}$ and $\overline{BC}$ where $\square ABCD$ and $\square EFCB$ share the base and point G where G is not a vertex. We claim that $\square ABCD = \square EFCB$.

Notice that $AD = BC = EF$. Since $AD + DE = AE$, $DE + EF = DF$, and $AD = EF$, we have $AE = DF$.

Consider $\triangle BAE$ and $\triangle CDF$: $BA = CD$, $AE = DF$, and $\angle BAE = \angle CDF$ by [1.29]. By [1.4], $\triangle BAE \cong \triangle CDF$, from which it follows that:

$$\begin{aligned} \triangle BAE &= \triangle CDF \\ \triangle BAE - \triangle DEG &= \triangle CDF - \triangle DEG \\ ADGB &= CGEF \\ ADGB + \triangle BGC &= CGEF + \triangle BGC \\ \square ABCD &= \square EFCB \end{aligned}$$

This proves case 3 and completes the proof. $\square$

CHAPTER 1. ANGLES, PARALLEL LINES, PARALLELOGRAMS

Proposition 1.36. *AREAS OF PARALLELOGRAMS II.*

Parallelograms on equal bases and on the same parallels are equal in area.

Proof. Construct $\square ADCB$ and $\square EHGF$ between $\overleftrightarrow{AH} \parallel \overleftrightarrow{BG}$ on bases $\overline{BC}$ and $\overline{FG}$ such that $\overline{BC} = \overline{FG}$. We claim that $\square ADCB = \square EHGF$.

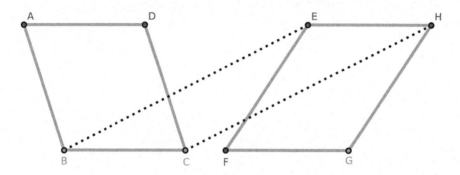

Figure 1.5.13: [1.36]

Construct $\overline{BE}$ and $\overline{CH}$. Since $\square EHGF$ is a parallelogram, $\overline{FG} = \overline{EH}$ [1.34]. Since $\overline{BC} = \overline{FG}$ by hypothesis, $\overline{BC} = \overline{EH}$. Given this equality and $\overline{BC} \parallel \overline{EH}$, by [1.33] $\overline{BE} = \overline{CH}$ and $\overline{BE} \parallel \overline{CH}$. It follows that $\square EHCB$ is a parallelogram.

By [1.35], $\square EHCB = \square EHGF$ and $\square EHCB = \square ADCB$. Therefore, $\square ADCB = \square EHGF$, which completes the proof. □

Proposition 1.37. *TRIANGLES OF EQUAL AREA I.*

Triangles which stand on the same base and in the same parallels are equal in area.

Proof. Construct $\triangle ABC$ and $\triangle DBC$ on base $\overline{BC}$ such that each triangle stands between parallels $\overleftrightarrow{AD}$ and $\overleftrightarrow{BC}$. We claim that $\triangle ABC = \triangle DBC$.

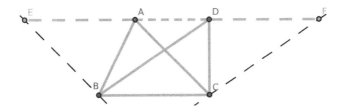

Figure 1.5.14: [1.37]

Construct $\overleftrightarrow{BE} \parallel \overleftrightarrow{AC}$ and $\overleftrightarrow{CF} \parallel \overleftrightarrow{BD}$. It follows that $\square AEBC$ and $\square DBCF$ are parallelograms. By [1.35], $\square AEBC = \square DBCF$.

Notice that $\frac{1}{2} \cdot \triangle ABC = \square AEBC$ because the diagonal AB bisects $\square AEBC$ [1.34, #1]. Similarly, $\frac{1}{2} \cdot \triangle DBC = \square DBCF$, and so

$$\frac{1}{2} \cdot \triangle ABC = \frac{1}{2} \cdot \triangle DBC$$
$$\triangle ABC = \triangle DBC$$

□

Exercises.

1. If two triangles of equal area stand on the same base but on opposite sides of the base, the segment connecting their vertices is bisected by the base or its extension. [See the final chapter for a solution.]

2. Construct a triangle equal in area to a given quadrilateral figure.

3. Construct a triangle equal in area to a given polygon.

4. Construct a rhombus equal in area to a given parallelogram and having a given side of the parallelogram as the base.

Proposition 1.38. *TRIANGLES OF EQUAL AREA II.*

Triangles which stand on equal bases and in the same parallels are equal in area.

Proof. Construct $\triangle ABC$ and $\triangle DEF$ between $\overleftrightarrow{BF}$ and $\overleftrightarrow{AD}$ such that $\overleftrightarrow{BF} \parallel \overleftrightarrow{AD}$ and $BC = EF$. We claim that $\triangle ABC = \triangle DEF$.

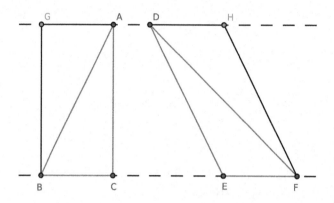

Figure 1.5.15: [1.38]

By [1.31], construct $\overline{BG}$ where G is a point on $\overleftrightarrow{AD}$ such that $\overline{BG} \parallel \overline{AC}$; similarly, construct $\overline{FH}$ where H is a point on $\overleftrightarrow{AD}$ such that $\overline{FH} \parallel \overline{DE}$. It follows that $\square GACB$ and $\square DHFE$ are parallelograms. Since $BC = EF$, by [1.36] $\square GACB = \square DHFE$.

Since $\overline{AB}$ bisects $\square GACB$ and $\overline{DF}$ bisects $\square DHFE$ by [1.34], $\triangle ABC = \triangle DEF$, completing the proof. $\square$

Exercises.

1. Every median of a triangle bisects the triangle. [See the final chapter for a solution.]

2. If two triangles have two sides of one respectively equal to two sides of the other and where the interior angles are supplemental, then their areas are equal.

3. If the base of a triangle is divided into any number of equal segments, then segments constructed from the vertex to the points of division divide the whole triangle into as many equal parts.

4. The diagonal of a parallelogram and segments from any point on the diagonal to the vertices through which the diagonal does not pass divide the parallelogram into four triangles which are equal (in a two-by-two fashion).

5. One diagonal of a quadrilateral bisects the other if and only if the diagonal also bisects the quadrilateral. [See the final chapter for a solution.]

6. If $\triangle ABC$ and $\triangle ABD$ each stand on the base AB and between the same parallels, and if a parallel to AB meets the sides AC and BC at the points E and F and meets the sides AD and BD at the points G, H, then $\overline{EF} = \overline{GH}$.

7. If instead of triangles on the same base we have triangles on equal bases and between the same parallels, the intercepts made by the sides of the triangles on any parallel to the bases are equal in length.

8. If the midpoints of any two sides of a triangle are joined, the triangle formed with the two half sides has an area equal to one-fourth of the whole.

9. The triangle whose vertices are the midpoints of two sides and any point on the remaining side has an area equal to one-fourth the area of that triangle.

10. Bisect a given triangle by a segment constructed from a given point in one of the sides.

11. Trisect a given triangle by three segments constructed from a given point within it.

12. Prove that any segment through the intersection of the diagonals of a parallelogram bisects the parallelogram.

13. The triangle formed by joining the midpoint of one of the non-parallel sides of a trapezoid to the endpoints of the opposite side is equal in area to half the area of the trapezoid. (Recall that a trapezoid is a quadrilaterals with two parallel sides and two non-parallel sides.)

Proposition 1.39. *TRIANGLES OF EQUAL AREA III.*

Triangles which are equal in area and stand on the same base and on the same side of the base also stand between the same parallels.

Proof. Suppose that $\triangle BAC$ and $\triangle BDC$ stand on the same base, BC, on the same side of BC, and that $\triangle BAC = \triangle BDC$; we claim $\triangle BAC$ and $\triangle BDC$ stand between $\overline{AD} \parallel \overline{BC}$.

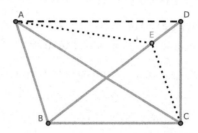

Figure 1.5.16: [1.39]

Construct $\overline{AD}$. Clearly, the triangles stand between $\overline{AD}$ and $\overline{BC}$. We need only prove that $\overline{AD} \parallel \overline{BC}$.

Suppose $\overline{AD} \nparallel \overline{BC}$ and that $\overline{AE} \parallel \overline{BC}$ where E is a point on $\overline{BD}$ other than D; construct $\overline{EC}$. Notice that $\triangle EDC > 0$ (if $\triangle EDC = 0$, then $E = D$, a contradiction).

Since the triangles $\triangle BEC$, $\triangle BAC$ stand on the same base BC and between the same parallels ($\overline{BC}$ and $\overline{AE}$), we find that $\triangle BEC = \triangle BAC$ [1.37]. By hypothesis, $\triangle BAC = \triangle BDC$. Therefore, $\triangle BDC = \triangle BEC$ [Axiom 1.1]. But $\triangle BDC = \triangle BEC + \triangle EDC$, and so $\triangle EDC = 0$ and $\triangle EDC > 0$. A similar contradiction results if we place E anywhere other than on D.

It follows that $\overline{AD} \parallel \overline{BC}$, which completes the proof. $\square$

1.5. BOOK I, PROPOSITIONS 27-48

Proposition 1.40. *TRIANGLES OF EQUAL AREA IV.*

Triangles which are equal in area and stand on equal bases on the same side of their bases stand between the same parallels.

Proof. Construct $\triangle ABC$ and $\triangle DEF$ on $\overline{BF}$ such that $BC = EF$, $\overline{CE} > 0$, $\triangle ABC = \triangle DEF$, and where each triangle stands on the same side of its base. Construct $\overline{AD}$. Clearly, $\triangle ABC$ and $\triangle DEF$ stand between $\overline{AD}$ and $\overline{BF}$; we claim that $\overline{AD} \parallel \overline{BF}$.

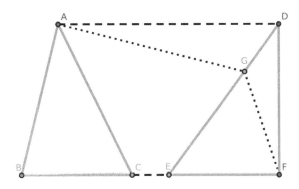

Figure 1.5.17: [1.40]

If $\overline{AD} \not\parallel \overline{BF}$, construct $\overline{AG}$ (where G is a point on $\overline{DE}$) such that $\overline{AG} \parallel \overline{BF}$. Also construct $\overline{FG}$. Notice that $\triangle DFG > 0$.

Consider the triangles $\triangle GEF$ and $\triangle ABC$: they stand on equal bases (BC, EF) and between the same parallels ($\overline{BF}$, $\overline{AG}$). By [1.38], $\triangle GEF = \triangle ABC$.

But $\triangle DEF = \triangle ABC$ by hypothesis, and so $\triangle DEF = \triangle GEF$. Since $\triangle DEF = \triangle GEF + \triangle DGF$, $\triangle DGF = 0$ and $\triangle DFG > 0$. A similar contradiction results if we place G anywhere other than on D.

It follows that $\overline{AD} \parallel \overline{BF}$, which completes the proof. □

Exercises.

1. Prove that triangles with equal bases and altitudes are equal in area. [See the final chapter for a solution.]

2. The segment joining the midpoints of two sides of a triangle is parallel to the base, and the medians from the endpoints of the base to these midpoints will each bisect the original triangle. Hence, the two triangles whose base is the third side and whose vertices are the points of bisection are equal in area. [See the final chapter for a solution.]

3. The parallel to any side of a triangle through the midpoint of another bisects the third.

4. The segments which connect the midpoints of the sides of a triangle divide the triangle into four congruent triangles. [See the final chapter for a solution.]

5. The segment which connects the midpoints of two sides of a triangle is equal in length to half the third side.

6. The midpoints of the four sides of a convex quadrilateral, taken in order, are the vertices of a parallelogram whose area is equal to half the area of the quadrilateral.

7. The sum of the two parallel sides of a trapezoid is double the length of the segment joining the midpoints of the two remaining sides.

8. The parallelogram formed by the segment which connects the midpoints of two sides of a triangle and any pair of parallels constructed through the same points to meet the third side is equal in area to half the area of the triangle.

9. The segment joining the midpoints of opposite sides of a quadrilateral and the segment joining the midpoints of its diagonals are concurrent.

Proposition 1.41. *PARALLELOGRAMS AND TRIANGLES.*

If a parallelogram and a triangle stand on the same base and between the same parallels, then the parallelogram is double the area of the triangle.

Proof. Construct $\square ABCD$ and $\triangle EBC$ on base $\overline{BC}$ and between $\overline{AE} \parallel \overline{BC}$. We claim that $\square ABCD = 2 \cdot \triangle EBC$.

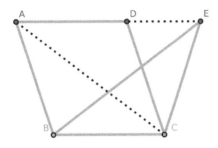

Figure 1.5.18: [1.41]

Construct $\overline{AC}$ and $\overline{DE}$. By [1.34], $\square ABCD = 2 \cdot \triangle ABC$; by [1.37], $\triangle ABC = \triangle EBC$. Therefore, $\square ABCD = 2 \cdot \triangle EBC$, which completes the proof. $\square$

Corollary. *1.41.1. If a triangle and a parallelogram have equal altitudes and if the base of the triangle is double of the base of the parallelogram, then their areas are equal.*

Corollary. *1.41.2. Suppose we have two triangles whose bases are two opposite sides of a parallelogram and which have any point between these sides as a common vertex. Then the sum of the areas of these triangles equals half the area of the parallelogram.*

CHAPTER 1. ANGLES, PARALLEL LINES, PARALLELOGRAMS

Proposition 1.42. *CONSTRUCTION OF PARALLELOGRAMS I.*

Given an arbitrary triangle and an arbitrary acute angle, it is possible to construct a parallelogram equal in area to the triangle which contains the given angle.

Proof. Construct $\triangle ABC$ and $\angle RDS$. We wish to construct $\square FGBE$ such that $\square FGBE = \triangle ABC$ and where $\square FGBE$ contains an angle equal in measure to $\angle RDS$.

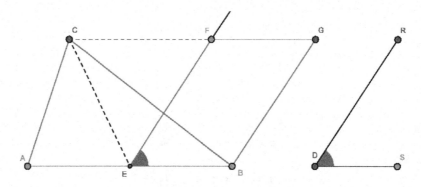

Figure 1.5.19: [1.42]

Bisect AB at E, and construct $\overline{EC}$. Construct $\angle BEF = \angle RDS$ [1.23], $\overline{CG} \parallel \overline{AB}$, and $\overline{BG} \parallel \overline{EF}$ [1.31].

Because $\overline{AE} = \overline{EB}$ by construction, $\triangle AEC = \triangle EBC$ by [1.38]. Therefore, $\triangle ABC = 2 \cdot \triangle EBC$. By [1.41], $\square FGBE = 2 \cdot \triangle EBC$. Therefore, $\square FGBE = \triangle ABC$.

Since $\square FGBE$ contains $\angle BEF$ where $\angle BEF = \angle RDS$, the proof is complete. □

1.5. BOOK I, PROPOSITIONS 27-48

Proposition 1.43. *COMPLEMENTS OF PARALLELOGRAMS.*

Parallel segments through any point in one of the diagonals of a parallelogram divides the parallelogram into four smaller parallelograms: the two through which the diagonal does not pass are called the *complements* of the other two parallelograms, and these complements are equal in area.

Proof. Construct $\square ABCD$ and diagonal $\overline{AC}$. Let K be any point on $\overline{AC}$ except A or C. Construct $\overline{GH}$ and $\overline{EF}$ through K such that $\overline{GH} \parallel \overline{CD}$ and $\overline{EF} \parallel \overline{AD}$. This divides $\square ABCD$ into four smaller parallelograms where $\square EBGK$, $\square HKFD$ are the *complements* of $\square AEKH$, $\square KGCF$. We claim that $\square EBGK = \square HKFD$.

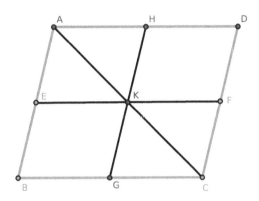

Figure 1.5.20: [1.43]

Because $\overline{AC}$ bisects the parallelograms $\square ABCD$, $\square AEKH$, and $\square KGCF$ by [1.34], we have $\triangle ADC = \triangle ABC$, $\triangle AHK = \triangle AEK$, and $\triangle KFC = \triangle KGC$. Therefore,

$$\begin{aligned}
\square EBGK &= \triangle ABC - \triangle AEK - \triangle KGC \\
&= \triangle ADC - \triangle AHK - \triangle KFC \\
&= \square HKFD
\end{aligned}$$

which completes the proof. □

Corollary. *1.43.1. If through some point K within parallelogram $\square ABCD$ we have constructed parallel segments to its sides in order to make the parallelograms $\square HDFK$, $\square EKGB$ equal in area, then K is a point on the diagonal $\overline{AC}$.*

Given [1.43], we find that $\square HDFK$, $\square EKGB$ equal in area if and only if K is a point on the diagonal $\overline{AC}$.

Corollary. *1.43.2. $\square AHGB = \square ADFE$ and $\square EFCB = \square HDCG$.*

102 CHAPTER 1. ANGLES, PARALLEL LINES, PARALLELOGRAMS

Exercises.

1. Prove [1.43, Cor. 1].

2. Prove [1.43, Cor. 2].

Proposition 1.44. *CONSTRUCTION OF PARALLELOGRAMS II.*

Given an arbitrary triangle, an arbitrary angle (acute, right, or obtuse), and an arbitrary segment, we can construct a parallelogram equal in area to the triangle which contains the given angle and has a side length equal to the given segment.

Proof. Construct $\angle RST$, $\triangle NPQ$, and $\overline{AB}$. We wish to construct $\square BALM$ on $\overline{AB}$ such that $\square BALM = \triangle NPQ$ and $\square BALM$ contains an angle equal to $\angle RST$.

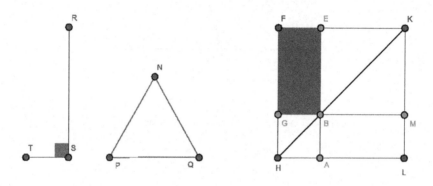

Figure 1.5.21: [1.44]

Construct the parallelogram $\square BEFG$ where $\square BEFG = \triangle NPQ$ [1.42], $\angle GBE = \angle RST$, and $\overline{AB} \oplus \overline{BE} = \overline{AE}$. Also construct segment $\overline{AH} \parallel \overline{BG}$ [1.31]. Extend $\overline{FG}$ to $\overline{FH}$, and construct $\overline{HB}$.

Since $\overline{AH} \parallel \overline{BG}$ and $\overline{BG} \parallel \overline{FE}$ by construction, $\overline{AH} \parallel \overline{FE}$ by [1.30]. Notice that $\overline{HF}$ intersects $\overline{AH}$ and $\overline{FE}$; therefore, $\angle AHF + \angle HFE =$ two right angles. It follows that $\angle BHG + \angle GFE <$ two right angles. By [1.3.3, Axiom 4], if we extend $\overline{HB}$ and $\overline{FE}$, they will intersect at some point K. Through K, construct $\overline{KL} \parallel \overline{AB}$ [1.31], extend $\overline{AH}$ to intersect $\overline{KL}$ at L, and extend $\overline{GB}$ to intersect $\overline{KL}$ at M. We claim that $\square BALM$ fulfills the required conditions.

Clearly, $\square BALM$ is constructed on $\overline{AB}$. By [1.43], $\square BMLA = \square FEBG$, and $\square FEBG = \triangle NPQ$ by construction; therefore $\square BALM = \triangle NPQ$. By [1.15], $\angle ABM = \angle EBG$, and $\angle EBG = \angle RST$ by construction; therefore, $\angle ABM = \angle RST$. This completes the proof. □

1.5. BOOK I, PROPOSITIONS 27-48

Proposition 1.45. *CONSTRUCTION OF PARALLELOGRAMS III.*

Given an arbitrary angle (acute, right, or obtuse) and an arbitrary polygon, we can construct a parallelogram equal in area to the polygon which contains an angle equal to the given angle.

Proof. Construct polygon $ABCD$ and $\angle LMN$. We wish to construct $\square FIKE$ such that it contains an angle equal to $\angle LMN$ and $\square FIKE = ABCD$.

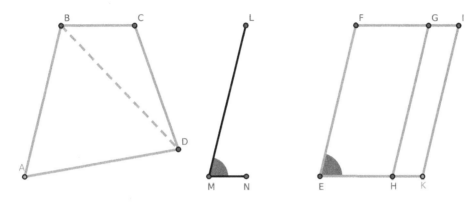

Figure 1.5.22: [1.45]

By [1.42], we may construct $\overline{BD}$ and $\square FGHE$ such that $\square FGHE = \triangle ABD$ where $\angle FEH = \angle LMN$.

On $\overline{GH}$, construct $\square GIKH$ such that $\square GIKH = \triangle BCD$ where $\angle GHK = \angle LMN$ [1.44]. (We may continue to this algorithm for any additional triangles that remain in $ABCD$. This allows us to claim that the proof which follows applies to any n–gon where $n < \infty$.) Upon completing this algorithm, we claim that $\square FEKI$ fulfills the required conditions.

Because $\angle GHK = \angle LMN = \angle FEH$ by construction, $\angle GHK = \angle FEH$. From this, we obtain

$$\angle GHK + \angle GHE = \angle FEH + \angle GHE$$

Since $\overline{HG} \parallel \overline{EF}$ and $\overline{EH}$ intersects them, the sum $\angle FEH + \angle GHE =$ two right angles [1.29]. Hence, $\angle GHK + \angle GHE =$ two right angles, and so $\overline{EH} \oplus \overline{HK} = \overline{EK}$ [1.14, Cor. 1]. Since $\overline{EH} \parallel \overline{FG}$ by construction, we now have $\overline{EK} \parallel \overline{FG}$.

Similarly to the above, because $\overline{GH}$ intersects the parallels $\overline{FG}$ and $\overline{EK}$, $\angle FGH = \angle GHK$ [Cor. 1.29.1], and so

$$\angle FGH + \angle HGI = \angle GHK + \angle HGI$$

Since $\overline{GI} \parallel \overline{HK}$ and $\overline{GH}$ intersects them, the sum $\angle GHK + \angle HGI =$ two right angles [1.29]. Hence, $\angle FGH + \angle HGI =$ two right angles, and $\overline{FG} \oplus \overline{GI} = \overline{FI}$.

Because $\square FEHG$ and $\square GHKI$ are parallelograms, $\overline{EF}$ and $\overline{KI}$ are each parallel to $\overline{GH}$; by [1.30], $\overline{EF} \parallel \overline{KI}$. Since $\angle GHK = \angle FGH$ by the above, by [Cor. 1.29.1], $\overline{EK} \parallel \overline{FI}$. Therefore, $\square FIKE$ is a parallelogram.

Since $\angle FEK = \angle LMN$, $\square FIKE$ contains an angle equal to a $\angle LMN$.

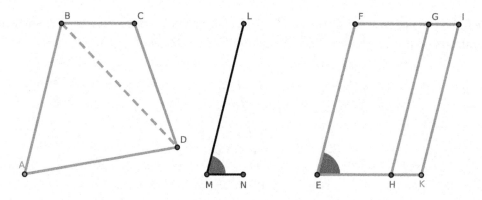

Figure 1.5.23: [1.45]

Because $\square FGHE = \triangle ABD$ by construction and $\square GIKH = \triangle BCD$,

$$\begin{aligned} ABCD &= \triangle ABD \oplus \triangle BCD \\ &= \square FGHE \oplus \square GIKH \\ &= \square FIKE \end{aligned}$$

which completes the proof. $\square$

Exercises.

1. Construct a rectangle equal to the sum of $2, 3, ..., n$ number of polygons.

2. Construct a rectangle equal in area to the difference in areas of two given figures.

Proposition 1.46. *CONSTRUCTION OF A SQUARE I.*

Given an arbitrary segment, we may construct a square on that segment.

Proof. Construct $\overline{AB}$; we wish to construct a square on $\overline{AB}$.

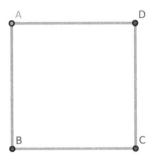

Figure 1.5.24: [1.46]

Construct $\overline{AD} \perp \overline{AB}$ [1.11] where $\overline{AD} = \overline{AB}$ [1.3]. Through D, construct $\overline{CD} \parallel \overline{AB}$ [1.31] where $\overline{AB} = \overline{CD}$, and through B construct $\overline{BC} \parallel \overline{AD}$. We claim that $\square ABCD$ is the required square.

By construction,
$$\overline{AD} = \overline{AB} = \overline{CD}$$

Because $\square ABCD$ is a parallelogram, $\overline{AD} = \overline{BC}$ [1.34]; hence, all four sides of $\square ABCD$ are equal. It follows that $\square ABCD$ is a rhombus and $\angle DAB$ is a right angle. By [Def. 1.30], $\square ABCD$ is a square. $\square$

Remark. [1.46] is a lemma to [1.47]. [2.14] offers a second method to construct a square.

Exercises.

1. Prove that two squares have equal side-lengths if and only if the squares are equal in area. [See the final chapter for a solution.]

2. Prove that the parallelograms about the diagonal of a square are squares.

3. If on the four sides of a square (or on the sides which are extended) points are taken which are equidistant from the four angles, then they will be the vertices of another square (and similarly for a regular pentagon, hexagon, etc.).

4. Divide a given square into five equal parts: specifically, four right triangles and a square.

5. Prove that the formula for the area of a rectangle is $A = bh$.

Proposition 1.47. *THE PYTHAGOREAN THEOREM (aka. THE GOUGU THEOREM).*

In a right triangle, the square on the side opposite the right angle (the hypotenuse) is equal in area to the sum of the areas of the squares on the remaining sides.

Proof. Construct right triangle $\triangle ABC$ where AB is the hypotenuse. We claim that

$$AB^2 = AC^2 + BC^2$$

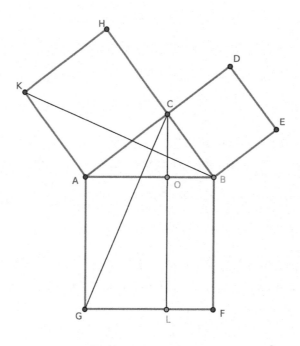

Figure 1.5.25: [1.47]

By [1.46], we may construct squares on sides AB, BC, and CA of $\triangle ABC$ as in Fig. 1.5.25. Construct $\overline{CL} \parallel \overline{AG}$ where L is a point on $\overline{GF}$. Also construct $\overline{CG}$ and $\overline{BK}$. Because both $\angle ACB$ and $\angle ACH$ are right angles by construction, the sum $\angle ACB + \angle ACH$ equals two right angles. Therefore $\overline{BC} \oplus \overline{CH} = \overline{BH}$ [1.14]. Similarly, $\overline{AC} \oplus \overline{CD} = \overline{AD}$.

Because $\angle BAG$ and $\angle CAK$ are each angles within a square, they are right angles. Hence,

$$\begin{aligned}
\angle BAG &= \angle CAK \\
\angle BAG + \angle BAC &= \angle CAK + \angle BAC \\
\angle CAG &= \angle KAB
\end{aligned}$$

1.5. BOOK I, PROPOSITIONS 27-48

Since $\square BAGF$ and $\square CHKA$ are squares, $BA = AG$ and $CA = AK$. Consider $\triangle CAG$ and $\triangle KAB$: since $CA = AK$, $BA = AG$, and $\angle CAG = \angle KAB$, by [1.4] $\triangle CAG \cong \triangle KAB$.

By [1.41], $\square AGLO = 2 \cdot \triangle CAG$ because they both stand on $\overline{AG}$ and stand between the parallels $\overline{AG}$ and $\overline{CL}$. Similarly, $\square CHKA = 2 \cdot \triangle KAB$ because they stand on $\overline{AK}$ and between $\overline{AK}$ and $\overline{BH}$. Since $\triangle CAG \cong \triangle KAB$, $\square AGLO = \square KACH$.

Similarly, it can be shown that $\square OLFB = \square DCBE$. Hence,

$$\begin{aligned} AB^2 &= \square AGFB \\ &= \square AGLO \oplus \square OLFB \\ &= \square KACH + \square DCBE \\ &= AC^2 + BC^2 \end{aligned}$$

which proves our claim. $\square$

Remark. [1.47] is a special case of [6.31].

Remark. [Cor. 10.28.1] describes a ratio which provide Pythagorean Triples[41] (three positive integers a, b, and c such that $a^2 + b^2 = c^2$): if a and b are positive integers and $b < a$, then Pythagorean Triples follow the ratio

$$ab : \frac{a^2 - b^2}{2} : \frac{a^2 + b^2}{2}$$

---

[41]https://en.wikipedia.org/wiki/Pythagorean_triple

Alternatively:

Proof. Construct the squares as in Fig. 1.5.26 such that $\triangle ACB$ is a right triangle where $\angle ACB$ is the right angle.

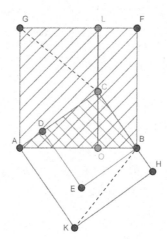

Figure 1.5.26: [1.47], alternate proof

Construct $\overline{CG}$ and $\overline{BK}$; through C construct $\overline{OL} \parallel \overline{AG}$. Notice that $\angle GAK = \angle GAC + \angle BAC + \angle BAK$ and that $\angle BAG$ and $\angle CAK$ are right angles. It follows that:

$$\begin{aligned} \angle BAG &= \angle CAK \\ \angle BAG - \angle BAC &= \angle CAK - \angle BAC \\ \angle CAG &= \angle BAK \end{aligned}$$

Consider $\triangle CAG$ and $\triangle BAK$: $CA = AK$, $AG = AB$, and $\angle CAG = \angle BAK$; by [1.4], $\triangle CAG \cong \triangle BAK$.

Applying [1.41], we find that $\square GAOL = \square AKHC$. Similarly, $\square LOBF = \square DEBC$. $\square$

Remark. The alternative proof is shorter since it's not necessary to prove that $\overline{AC}$ and $\overline{CD}$ form one segment. Similarly, the proposition may be proven by taking any of the eight figures formed by turning the squares in all possible directions. Another simplification of the proof can be obtained by considering that the point A is such that one of the triangles $\triangle CAG$ or $\triangle BAK$ can be turned round it in its own plane until it coincides with the other; hence, they are congruent.

Exercises.

1. Prove that the square on $\overline{AC}$ is equal in area to the rectangle $\square\overline{AB} \cdot \overline{AO}$, and the square on $\square\overline{BC} = \square\overline{AB} \cdot \overline{BO}$. (Note: $\square\overline{AB} \cdot \overline{AO}$ denotes the rectangle formed by the segments $\overline{AB}$ and $\overline{AO}$.)

2. Prove that the square on $\square CO = \square AO \cdot OB$.

3. Prove that $AC^2 - BC^2 = AO^2 - BO^2$

4. Find a segment whose square is equal to the sum of the areas of two given squares. [See the final chapter for a solution.]

5. Given the base of a triangle and the difference of the squares of its sides, the locus of its vertex is a segment perpendicular to the base.

6. In Fig. 1.5.25, prove that $\overline{BK} \perp \overline{CG}$. $\overline{BK}$ and $\overline{CG}$ are transverse segments.

7. In Fig. 1.5.25: if $\overline{EG}$ is constructed, prove that $\overline{EG}^2 = \overline{AC}^2 + 4 \cdot \overline{BC}^2$.

8. The square constructed on the sum of the sides of a right triangle exceeds the square on the hypotenuse by four times the area of the triangle (see [1.46, #3]). More generally, if the vertical angle of a triangle is equal to the angle of a regular polygon of n sides, then the regular polygon of n sides, constructed on a segment equal to the sum of its sides exceeds the area of the regular polygon of n sides constructed on the base by n times the area of the triangle.

9. If $\overline{AC}$ and $\overline{BK}$ intersect at P and a segment is constructed through P which is parallel to $\overline{BC}$, meeting $\overline{AB}$ at Q, then $\overline{CP} = \overline{PQ}$.

10. Prove that each of the triangles $\triangle AGK$ and $\triangle BEF$ formed by joining adjacent corners of the squares in [1.47] is equal in area to $\triangle ABC$. [See the final chapter for a solution.]

11. Find a segment whose square is equal to the difference of the squares on two segments.

12. The square on the difference of the sides $\overline{AC}, \overline{CB}$ is less than the square on the hypotenuse by four times the area of the triangle.

13. If $\overline{AE}$ is connected, then the segments $\overline{AE}, \overline{BK}, \overline{CL}$ are concurrent.

14. In an equilateral triangle, three times the square on any side is equal to four times the square on the perpendicular to it from the opposite vertex.

15. We construct the square $\square BEFG$ on $\overline{BE}$, a part of the side $\overline{BC}$ of a square $\square ABCD$, having its side $\overline{BG}$ in the continuation of $\overline{AB}$. Divide the figure $AGFECD$ into three parts which will form a square.

16. Four times the sum of the squares on the medians which bisect the sides of a right triangle is equal to five times the square on the hypotenuse.

17. If perpendiculars fall on the sides of a polygon from any point and if we divide each side into two segments, then the sum of the squares on one set of alternate segments is equal to the sum of the squares on the remaining set.

18. The sum of the squares on segments constructed from any point to one pair of opposite angles of a rectangle is equal to the sum of the squares on the segments from the same point to the remaining pair.

19. Divide the hypotenuse of a right triangle into two parts such that the difference between their squares equals the square on one of the sides.

20. From the endpoints of the base of a triangle, let altitudes fall on the opposite sides. Prove that the sum of the rectangles contained by the sides and their lower segments is equal to the square on the base.

Proposition 1.48. *THE CONVERSE OF THE PYTHAGOREAN/GOUGU THEOREM.*

Construct squares on all sides of a triangle. If the square on the hypotenuse is equal in area to the sum of the areas of the squares on the remaining sides, then the angle opposite to the longest side is a right angle.

Proof. Construct $\triangle ABC$ such that AB is the longest side and

$$AB^2 = AC^2 + BC^2$$

We claim that $\angle ACB$ is a right angle.

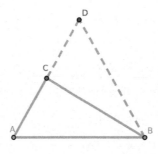

Figure 1.5.27: [1.48]

Construct $\overline{CD}$ such that $\overline{CD} = \overline{CA}$ [1.3] and $\overline{CD} \perp \overline{CB}$ [1.11]. Construct $\overline{BD}$, and consider $\triangle BCD$: $\angle BCD$ is a right angle by construction. $AC = CD$ implies that $AC^2 = CD^2$, and so

$$AC^2 + CB^2 = CD^2 + CB^2$$

1.5. BOOK I, PROPOSITIONS 27-48

By [1.47] $CD^2 + CB^2 = BD^2$; by hypothesis, $AC^2 + CB^2 = AB^2$. Hence $AB^2 = BD^2$; it follows that $AB = BD$ [1.46, #1].

Consider $\triangle ACB$ and $\triangle DCB$: $AB = DB$, $AC = CD$ by construction, and each shares the side CD. By [1.8], $\triangle ACB \cong \triangle DCB$, and so $\angle ACB = \angle DCB$. Since $\angle DCB$ is a right angle by construction, $\angle ACB$ is also a right angle. $\square$

An alternate proof by contradiction:

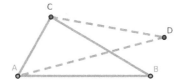

Figure 1.5.28: [1.48], alternative proof

Proof. Construct $\triangle ABC$ such that $AC^2 + BC^2 = AB^2$. If $CD \not\perp CA$, construct $\overline{CD} \perp \overline{CA}$ such that $\overline{CD} = \overline{CB}$. Construct $\overline{AD}$.

Consider $\triangle ABC$ and $\triangle ADC$: $\overline{CD} = \overline{CB}$, the triangles share side AC, and as in the above proof, it can be shown that $AD = AB$. This contradicts [1.7]; it follows that, $\angle ACB$ is a right angle. $\square$

Corollary. *1.48.1 Let a, b, and c be sides of $\triangle A$ where c is the longest side. $\triangle A$ is a right triangle if and only if $a^2 + b^2 = c^2$.*

Exam questions on chapter 1.

1. What is geometry?

2. What is geometric object?

3. Name the primary concepts of geometry. (Ans. Points, lines, surfaces, and solids.)

4. What kinds of lines exist in geometry? (Ans. Straight and curved.)

5. How is a straight line constructed? (Ans. By connecting any two collinear points.)

6. How is a curved line constructed? (Ans. By connecting any three non-collinear points.)

7. How may surfaces be divided? (Ans. Into planes and curved surfaces.)

8. How may a plane surface be constructed?

9. Why does a point have no dimensions?

10. Does a line have either width nor thickness?

11. How many dimensions does a surface possess?

12. What is plane geometry?

13. What portion of plane geometry forms the subject of this chapter?

14. What is the subject-matter of the remaining chapters?

15. How is a proposition proved indirectly?

16. What is meant by the inverse of a proposition?

17. What proposition is an instance of the Rule of Symmetry?

18. What are congruent figures?

19. What is another way to describe congruent figures? (Ans. They are identically equal.)

20. Mention all the instances of equality which are not congruence that occur in chapter 1.

21. What is the difference between the symbols denoting congruence and equality?

22. Define adjacent, exterior, interior, and alternate angles.

23. What is meant by the projection of one line on another?

24. What are meant by the medians of a triangle?

25. What is meant by the third diagonal of a quadrilateral?

26. State some propositions in chapter 1 which are particular cases of more general ones that follow.

27. What is the sum of all the exterior angles of any polygon equal to?

28. How many conditions must be given in order to construct a triangle? (Ans. Three; such as the three sides, or two sides and an angle, etc.)

Chapter 1 exercises.

1. Suppose $\triangle_1$ and $\triangle_2$ are triangles such that:

   (a) $\triangle_1$ is constructed within $\triangle_2$

   (b) each side of $\triangle_2$ passes through one vertex of $\triangle_1$

   (c) each side of $\triangle_2$ is parallel to its opposite side in $\triangle_1$

   We claim that $\triangle_2 = 4 \cdot \triangle_1$. [See the final chapter for a solution.]

2. The three altitudes of the first triangle in #1 are the altitudes at the midpoints of the sides of the second triangle.

3. Through a given point, construct a line so that the portion intercepted by the segments of a given angle are bisected at the point.

4. The three medians of a triangle are concurrent. (Note: we are proving the existence of the **centroid** of a triangle. Students are encouraged to use Ceva's Theorem, not found in Euclid, to solve this problem. Students who seek a challenge should attempt this problem without using Ceva's Theorem.)

5. Construct a triangle given two sides and the median of the third side.

6. Let P =the perimeter of a triangle and S =the sum of the lengths of a triangle's medians. Prove that $\frac{3}{4} \cdot P < S < P$.

7. Construct a triangle given a side and the two medians of the remaining sides.

8. Construct a triangle given the three medians. [See the final chapter for a solution.]

9. The angle included between the perpendicular from the vertical angle of a triangle on the base and the bisector of the vertical angle is equal to half the difference of the base angles.

10. Find in two parallels two points which are equidistant from a given point and whose connecting line is parallel to a given line.

11. Construct a parallelogram given two diagonals and a side.

12. The shortest median of a triangle corresponds to the largest side.

13. Find in two parallels two points standing opposite a right angle at a given point and which are equally distant from it.

14. The sum of the distances of any point in the base of an isosceles triangle from the equal sides is equal to the distance of either endpoint of the base from the opposite side.

15. The three perpendiculars at the midpoints of the sides of a triangle are concurrent. Hence, prove that perpendiculars from the vertices on the opposite sides are concurrent.

16. Inscribe a lozenge in a triangle having for an angle one angle of the triangle. [See the final chapter for a solution.]

17. Inscribe a square in a triangle having its base on a side of the triangle.

18. Find the locus of a point, the sum or the difference of whose distance from two fixed lines is equal to a given length.

19. The sum of the perpendiculars from any point in the interior of an equilateral triangle is equal to the perpendicular from any vertex on the opposite side.

20. Find a point in one of the sides of a triangle such that the sum of the intercepts made by the other sides on parallels constructed from the same point to these sides are equal to a given length.

21. If two angles exist such that their segments are respectively parallel, then their bisectors are either parallel or perpendicular.

22. Inscribe in a given triangle a parallelogram whose diagonals intersect at a given point.

23. Construct a quadrilateral where the four sides and the position of the midpoints of two opposite sides are given.

24. The bases of two or more triangles having a common vertex are given, both in magnitude and position, and the sum of the areas is given. Prove that the locus of the vertex is a straight line.

25. If the sum of the perpendiculars from a given point on the sides of a given polygon is given, then the locus of the point is a straight line.

26. If $\triangle ABC$ is an isosceles triangle whose equal sides are AB, AC and if $B'C'$ is any secant cutting the equal sides at B', C', such that $AB' + AC' = AB + AC$, prove that $B'C' > BC$.

27. If A, B are two given points and P is a point on a given line L, prove that the difference between AP and PB is a maximum when L bisects the angle $\angle APB$. Show that their sum is a minimum if it bisects the supplement.

28. Bisect a quadrilateral by a segment constructed from one of its vertices.

29. If $\overleftrightarrow{AD}$ and $\overleftrightarrow{BC}$ are two parallel lines cut obliquely by $\overleftrightarrow{AB}$ and perpendicularly by $\overleftrightarrow{AC}$, and between these lines we construct BED, cutting $\overleftrightarrow{AC}$ at point E such that $\overline{ED} = 2 \cdot \overline{AB}$, prove that the angle $\angle DBC = \frac{1}{3} \cdot \angle ABC$.

30. If O is the point of concurrence of the bisectors of the angles of the triangle $\triangle ABC$, if $\overline{AO}$ is extended to intersect $\overline{BC}$ at D, and if $\overline{OE}$ is constructed from O such that $\overline{OE} \perp \overline{BC}$, prove that the $\angle BOD = \angle COE$.

31. The angle made by the bisectors of two consecutive angles of a convex quadrilateral is equal to half the sum of the remaining angles; the angle made by the bisectors of two opposite angles is equal to half the difference of the two other angles.

32. If in the construction of [1.47] we join EF, KG, then $EF^2 + KG^2 = 5 \cdot AB^2$.

33. Given the midpoints of the sides of a convex polygon of an odd number of sides, construct the polygon.

34. Trisect a quadrilateral by lines constructed from one of its angles.

35. Given the base of a triangle in magnitude and position and the sum of the sides, prove that the perpendicular at either endpoint of the base to the adjacent side and the external bisector of the vertical angle meet on a given line perpendicular to the base.

36. The bisectors of the angles of a convex quadrilateral form a quadrilateral whose opposite angles are supplemental. If the first quadrilateral is a parallelogram, the second is a rectangle; if the first is a rectangle, the second is a square.

37. Suppose that the midpoints of the sides AB, BC, CA of a triangle are respectively D, E, F and that $DG \parallel BF$ and intersects EF. Prove that the sides of the triangle $\triangle DCG$ are respectively equal to the three medians of the triangle $\triangle ABC$.

38. Find the path of a pool ball started from a given point which, after being reflected from the four sides of the table, will pass through another given point. (Assume that the ball does not enter a pocket.)

39. If two segments which bisect two angles of a triangle and are terminated by the opposite sides are equal in length, prove that the triangle is isosceles.

40. If a square is inscribed in a triangle, the rectangle under its side and the sum of the base and altitude is equal to twice the area of the triangle.

41. If AB, AC are equal sides of an isosceles triangle and if $BD \perp AC$, prove that $BC^2 = 2 \cdot AC \cdot CD$.

42. Given the base of a triangle, the difference of the base angles, and the sum or difference of the sides, construct it.

43. Given the base of a triangle, the median that bisects the base, and the area, construct it.

44. If the diagonals AC and BD of a quadrilateral $ABCD$ intersect at E and are bisected at the points F and G, then

$$4 \cdot \triangle EFG = (AEB + ECD) - (AED + EBC)$$

45. If squares are constructed on the sides of any triangle, the lines of connection of the adjacent corners are respectively:

(a) the doubles of the medians of the triangle;

(b) perpendicular to them.

Chapter 2

Rectangles

Chapter 2 proves a number of propositions that are familiar in the form of algebraic equations. Algebra as we know it had not been developed when Euclid wrote "The Elements", and so the results are more of historical importance than practical use (except when they are used in subsequent propositions). This is why Book II appears in truncated form.

If definitions, postulates, or axioms from chapter 1 are used, they generally won't be cited.

2.1 Definitions

1. If $\overline{AB}$ contains point C, then C is the *point of division* between $\overline{AC}$ and $\overline{CB}$. (Notice that the midpoint of $\overline{AB}$ is a special case of all such points of division.)

2. If $\overline{AB}$ is extended to point C, then point C is called a *point of external division*.

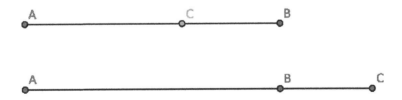

Figure 2.1.1: [Def. 2.1] above, [Def 2.2] below

3. A parallelogram whose angles are right angles is called a *rectangle*.

Figure 2.1.2: [Def. 2.3, 2.4, and 2.5]

4. A rectangle is said to be contained by any two adjacent sides: thus, □$ABCD$ is contained by sides AB and AD, or by sides AB and BC, etc.

5. The rectangle contained by two separate adjacent sides (such as AB and AD above) is the parallelogram formed by constructing a perpendicular to AB at B which is equal in length to AD and then constructing parallels.

The area of the rectangle is written $AB \cdot AD$.

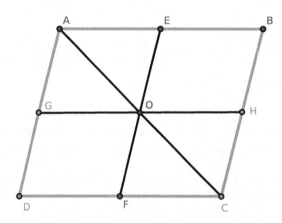

Figure 2.1.3: [Def. 2.6]

6. In any parallelogram, a figure which is composed of either of the parallelograms about a diagonal and the two complements is called a *gnomon* [see also 1.43]. If in Fig. 2.1.3 we remove either of the parallelograms □$AGDE$ or □$OFCH$ (but not both) from the parallelogram □$ADCB$, the remaining object is a gnomon.

7. A segment divided as in [2.11] is said to be divided in "extreme and mean ratio."

2.2 Axioms

1. A *semicircle* (half-circle) may be constructed given only a center point and a radius.

2.3 Propositions from Book II

Proposition 2.1. *Suppose that two segments ($\overline{AB}$, $\overline{BD}$) which intersect at one point (B) are constructed such that one segment ($\overline{BD}$) is divided into an arbitrary but finite number of segments ($\overline{BC}$, $\overline{CE}$, $\overline{EF}$, $\overline{FD}$). Then the rectangle contained by the two segments $\overline{AB}$, $\overline{BD}$ is equal in area to the sum of the areas of the rectangles contained by $\overline{AB}$ and the subsegments of the divided segment.*

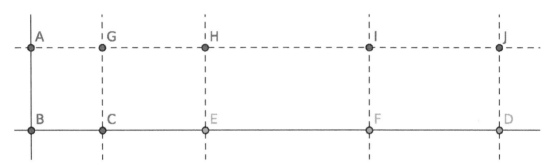

Figure 2.3.1: [2.1]

Corollary. *2.1.1. Algebraically, [2.1] states that the area*

$$AB \cdot BD = AB \cdot BC + AB \cdot CE + AB \cdot EF + AB \cdot FD$$

More generally, if $y = y_1 + y_2 + ... + y_n$, then

$$\begin{aligned} xy &= x(y_1 + ... + y_n) \\ &= xy_1 + xy_2 + ... + xy_n \end{aligned}$$

[Cor. 2.1.1] restates the Distributive Property from [1.3.2] Congruence Axioms.

Corollary. *2.1.2. The rectangle contained by a segment and the difference of two other segments equals the difference of the rectangles contained by the segment and each of the others.*

Corollary. *2.1.3. The area of a triangle is equal to half the rectangle contained by its base and perpendicular.*

Exercises.

1. Prove [Cor 2.1.1].

2. Prove [Cor. 2.1.2].

3. Prove [Cor. 2.1.3].

Proposition 2.2. *If a segment $(\overline{AB})$ is divided into any two subsegments at a point (C), then the square on the entire segment is equal in area to the sum of the areas of the rectangles contained by the whole and each of the subsegments $(\overline{AC}, \overline{CB})$.*

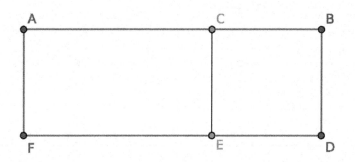

Figure 2.3.2: [2.2]

Corollary. 2.2.1. *Algebraically, [2.2] is a special case of [2.1] when $n = 2$. Specifically, it states that*
$$AF \cdot FD = AF \cdot FE + AF \cdot ED$$
or: if $y = y_1 + y_2$, then
$$\begin{aligned} xy &= x(y_1 + y_2) \\ &= xy_1 + xy_2 \end{aligned}$$

Exercise.

1. Prove [Cor. 1.2.1].

2.3. PROPOSITIONS FROM BOOK II

Proposition 2.3. *If a segment ($\overline{AB}$) is divided into two subsegments (at C), the rectangle contained by the whole segment and either subsegment (CB or CF) is equal to the square on that segment together with the rectangle contained by each of the segments.*

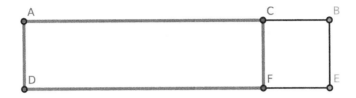

Figure 2.3.3: [2.3]

Corollary. 2.3.1. Algebraically, [2.3] states that if $x = y + z$, then

$$\begin{aligned} xy &= (y+z)y \\ &= y^2 + yz \end{aligned}$$

Exercise.

1. Prove [Cor. 1.3.1].

Proposition 2.4. *If a segment ($\overline{AB}$) is divided into any two parts (at C), the square on the whole segment is equal in area to the sum of the areas of the squares on the subsegments ($\overline{AC}$, $\overline{CB}$) together with twice the area of their rectangle.*

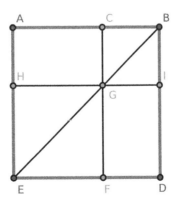

Figure 2.3.4: [2.4]

Corollary. *2.4.1. Algebraically, [2.4] states that if $x = y + z$, then*

$$\begin{aligned} x^2 &= (y+z)^2 \\ &= y^2 + 2yz + z^2 \end{aligned}$$

where $AC = y$, $CB = z$, and $AH \cdot GH = GF \cdot FI = yz$.

Corollary. *2.4.2. The parallelograms about the diagonal of a square are squares.*

Corollary. *2.4.3. If a segment is divided into any number of subsegments, the square on the whole is equal in area to the sum of the areas of the squares on all the subsegments, together with twice the sum of the areas of the rectangles contained by the several distinct pairs of subsegments.*

Corollary. *2.4.4. The square on a segment is equal in area to four times the square on its half.*

Exercises.

1. Prove [2.4] by using [2.2] and [2.3].

2. If from the right angle of a right triangle a perpendicular falls on the hypotenuse, its square equals the area of the rectangle contained by the segments of the hypotenuse. [See the final chapter for a solution.]

3. If from the hypotenuse of a right triangle subsegments are cut off equal to the adjacent sides, prove that the square on the middle segment is equal in area to twice the area of rectangle contained by the segments at either end.

4. In any right triangle, the square on the sum of the hypotenuse and perpendicular from the right angle on the hypotenuse exceeds the square on the sum of the sides by the square on the perpendicular.

5. The square on the perimeter of a right-angled triangle equals twice the rectangle contained by the sum of the hypotenuse and one side and the sum of the hypotenuse and the other side.

6. Prove [Cor. 2.4.1].

7. Prove [Cor. 2.4.2].

8. Prove [Cor. 2.4.3].

9. Prove [Cor. 2.4.4]. [See the final chapter for a solution.]

2.3. PROPOSITIONS FROM BOOK II

Proposition 2.5. *If a segment ($\overline{AB}$) is divided into two equal parts (at C) and also into two unequal parts (at D), the rectangle contained by the unequal parts ($\overline{AD}$, $\overline{DB}$) together with the square on the part between the points of section ($\overline{CD}$) is equal in area to the square on half the line.*

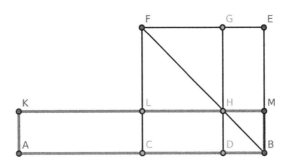

Figure 2.3.5: [2.5]

Corollary. *2.5.1. Algebraically, [2.5] states*[1]

$$xy = \frac{(x+y)^2}{2} + \frac{(x-y)^2}{2}$$

This may also be expressed as $\overline{AD} \cdot \overline{DB} + (\overline{CD})^2 = (\overline{AC})^2 = (\overline{CB})^2$.

Corollary. *2.5.2. The rectangle $\overline{AD} \cdot \overline{DB}$ is the rectangle contained by the sum of the segments $\overline{AC}$, $\overline{CD}$ and their difference, and we have proved it equal to the difference between the square on $\overline{AC}$ and the square on $\overline{CD}$. Hence the difference of the squares on two segments is equal to the rectangle contained by their sum and their difference.*

Corollary. *2.5.3. The perimeter of the rectangle $\overline{AH} = 2 \cdot \overline{AB}$, and is therefore independent of the position of the point D on $\overline{AB}$. The area of the same rectangle is less than the square on half the segment by the square on the subsegment between D and the midpoint of the line; therefore, when D is the midpoint, the rectangle will have the maximum area. Hence, of all rectangles having the same perimeter, the square has the greatest area.*

Exercises.

1. Divide a given segment so that the rectangle contained by its parts has a maximum area.

2. Divide a given segment so that the rectangle contained by its subsegments is equal to a given square, not exceeding the square on half the given line.

---

[1] http://aleph0.clarku.edu/~djoyce/java/elements/bookII/propII5.html

3. The rectangle contained by the sum and the difference of two sides of a triangle is equal to the rectangle contained by the base and the difference of the segments of the base made by the perpendicular from the vertex.

4. The difference of the sides of a triangle is less than the difference of the segments of the base made by the perpendicular from the vertex.

5. The difference between the square on one of the equal sides of an isosceles triangle and the square on any segment constructed from the vertex to a point in the base is equal to the rectangle contained by the segments of the base.

6. The square on either side of a right triangle is equal to the rectangle contained by the sum and the difference of the hypotenuse and the other side.

7. Prove [Cor. 2.5.1].

8. Prove [Cor. 2.5.2].

9. Prove [Cor. 2.5.3].

2.3. PROPOSITIONS FROM BOOK II

Proposition 2.6. *If a segment ($\overline{AB}$) is bisected (at C) and extended to a segment ($\overline{BD}$), the rectangle contained by the segments ($\overline{AD}, \overline{BD}$) made by the endpoint of the second segment (D) together with the square on half of the segment ($\overline{CB}$) equals the square on the segment between the midpoint and the endpoint of the second segment.*

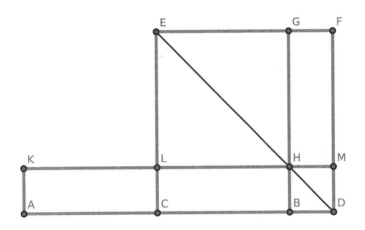

Figure 2.3.6: [2.6]

Corollary. *2.6.1. Algebraically, [2.6] states that*[2]

$$x(x-b) = (x - \frac{b}{2})^2 - (\frac{b}{2})^2$$

This may also be expressed as $\overline{AD} \cdot \overline{BD} + (\overline{CB})^2 = (\overline{CD})^2$.

Exercises.

1. Show that [2.6] is reduced to [2.5] by extending the line in the opposite direction.

2. Divide a given segment externally so that the rectangle contained by its subsegments is equal to the square on a given line.

3. Given the difference of two segments and the rectangle contained by them, find the subsegments.

4. The rectangle contained by any two segments equals the square on half the sum minus the square on half the difference.

5. Given the sum or the difference of two lines and the difference of their squares, find the lines.

6. If from the vertex C of an isosceles triangle a segment $\overline{CD}$ is constructed to any point in the extended base, prove that $(\overline{CD})^2 - (\overline{CB})^2 = \overline{AD} \cdot \overline{DB}$.

---

[2] http://aleph0.clarku.edu/~djoyce/java/elements/bookII/propII6.html

7. Give a common statement which will include [2.5] and [2.6]. [See the final chapter for a solution.]

8. Prove [Cor. 2.6.1].

Proposition 2.7. *If a segment $(\overline{AB})$ is divided into any two parts (at C), the sum of the areas of the squares on the whole segment (AB) and either subsegment $(\overline{CB})$ equals twice the rectangle $(\overline{AB}, \overline{CB})$ contained by the whole segment and that subsegment, together with the square on the remaining segment.*

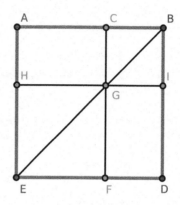

Figure 2.3.7: [2.7]

Corollary. *2.7.1. Algebraically, [2.7] states that if $x = y + z$, then*[3]

$$\begin{aligned} x^2 + z^2 &= (y+z)^2 + z^2 \\ &= y^2 + 2yz + 2z^2 \\ &= y^2 + 2z(y+z) \\ &= y^2 + 2xz \end{aligned}$$

Or,
$$(\overline{AB})^2 + (\overline{BC})^2 = 2 \cdot \overline{AB} \cdot \overline{BC} + (\overline{AC})^2$$

Equivalently, this result can be stated as $x^2 + z^2 = 2xz + (x-z)^2$.

Corollary. *2.7.2. Comparison of [2.4] and [2.7]:*

[2.4]: square on sum = sum of the areas of squares + twice rectangle

[2.7]: square on difference = sum of the areas of squares − twice rectangle

---

[3]http://aleph0.clarku.edu/~djoyce/java/elements/bookII/propII7.html

2.3. PROPOSITIONS FROM BOOK II

Proposition 2.8. *If a segment $(\overline{AB})$ is cut arbitrarily (at C), then four times the area of the rectangle contained by the whole and one of the segments $(\overline{AB}, \overline{BC})$ plus the area of the square on the remaining segment $(\overline{AC})$ equals the area of the square constructed on the whole and the aforesaid segment constructed as on one segment $(\overline{AB} \oplus \overline{BC})$.*[4]

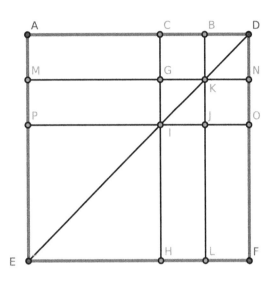

Figure 2.3.8: [2.8]

Corollary. 2.8.1. *Algebraically, [2.8] states that if $x = y + z$, then*

$$\begin{aligned}
(x+y)^2 &= x^2 + 2xy + y^2 \\
&= (y+z)^2 + 2(y+z)y + y^2 \\
&= y^2 + 2yz + z^2 + 2y^2 + 2yz + y^2 \\
&= 4y^2 + 4yz + z^2 \\
&= 4y(y+z) + z^2 \\
&= 4xy + z^2 \\
&= 4xy + (x-y)^2
\end{aligned}$$

Exercises.

1. In [1.47], if EF, GK are joined, prove that $(\overline{EF})^2 - (\overline{CO})^2 = (\overline{AB} + \overline{BO})^2$.

2. In [1.47], prove that $(\overline{GK})^2 - (\overline{EF})^2 = 3 \cdot \overline{AB} \cdot (\overline{AO} - \overline{BO})$.

3. Given that the difference of two segments equals R and the area of their rectangle equals $4R^2$, find the segments.

---

[4]http://aleph0.clarku.edu/~djoyce/java/elements/bookII/propII8.html

Proposition 2.9. *If a segment $\overline{AB}$ is bisected (at C) and divided into two unequal segments (at D), the area of the squares on the unequal subsegments ($\overline{AD}$, $\overline{DB}$) is double the area of the squares on half the line ($\overline{AC}$) and on the segment ($\overline{CD}$) between the points of section.*

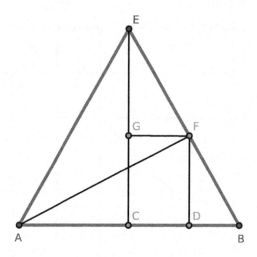

Figure 2.3.9: [2.9]

Corollary. *Algebraically, [2.9] states that*

$$(y+z)^2 + (y-z)^2 = 2(y^2 + z^2)$$

Exercises.

1. The sum of the squares on the subsegments of a larger segment of fixed length is a minimum when it is bisected.

2. Divide a given segment internally so that the sum of the areas of the squares on the subsegments equals the area of a given square and state the limitation to its possibility.

3. If a segment $\overline{AB}$ is bisected at C and divided unequally in D, then $(\overline{AD})^2 + (\overline{DB})^2 = 2 \cdot \overline{AD} \cdot \overline{DB} + 4 \cdot (\overline{CD})^2$.

4. Twice the area of a square on the segment joining any point in the hypotenuse of a right isosceles triangle to the vertex is equal to the sum of the areas of the squares on the segments of the hypotenuse.

5. If a segment is divided into any number of subsegments, the continued product of all the parts is a maximum and the sum of the areas of their squares is a minimum when all the parts are equal.

2.3. PROPOSITIONS FROM BOOK II

Proposition 2.10. *If a segment $(\overline{AB})$ is bisected (at C) and is extended to a segment $(\overline{AD})$, the sum of the areas of the squares on the segments $(\overline{AD}, \overline{DB})$ made by the endpoint (D) is equal to twice the area of the square on half the segment and twice the square on the segment between the points of that section.*

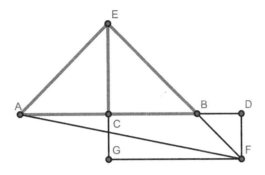

Figure 2.3.10: [2.10]

Corollary. *2.10.1. Algebraically, [2.10] states the same result as Proposition 2.9:*

$$(y+z)^2 + (y-z)^2 = 2(y+z)^2$$

Corollary. *2.10.2. The square on the sum of any two segments plus the square on their difference equals twice the area of the sum of their squares.*

Corollary. *2.10.3. The sum of the area of the squares on any two segments is equal to twice the area of the square on half the sum plus twice the square on half the difference of the lines.*

Corollary. *2.10.4. If a segment is cut into two unequal subsegments and also into two equal subsegments, the sum of the area of the squares on the two unequal subsegments exceeds the sum of the areas of the squares on the two equal subsegments by the sum of the areas of the squares of the two differences between the equal and unequal subsegments.*

Exercises.

1. Given the sum or the difference of any two segments and the sum of the areas of their squares, find the segments.

2. Consider $\triangle ABC$: the sum of the areas of the squares on two sides AC, CB is equal to twice the area of the square on half the base AB and twice the square on the median which bisects AB.

3. If the base of a triangle is given both in magnitude and position and the sum of the areas of the squares on the sides in magnitude, the locus of the vertex is a circle.

4. Consider $\triangle ABC$: if a point D on the base BC exists such that $(\overline{BA})^2 + (\overline{BD})^2 = (\overline{CA})^2 + (\overline{CD})^2$, prove that the midpoint of AD is equally distant from both B and C.

5. Prove [Cor. 2.10.1].

6. Prove [Cor. 2.10.2].

7. Prove [Cor. 2.10.3].

8. Prove [Cor. 2.10.4].

Proposition 2.11. *It is possible to divide a given segment ($\overline{AB}$) into two segments (at H) such that the rectangle ($\overline{AB}$, $\overline{BH}$) contained by the segment and its subsegment is equal in area to the square on the remaining segment ($\overline{AH}$).*

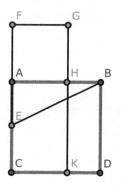

Figure 2.3.11: [2.11]

Corollary. *2.11.1. Algebraically, [2.11] solves the equation $\overline{AB} \cdot \overline{BH} = (\overline{AH})^2$, or $a(a - x) = x^2$. Specifically,*

$$\begin{aligned} a(a - x) &= x^2 \\ a^2 - ax &= x^2 \\ x^2 + ax - a^2 &= 0 \\ x &= -\tfrac{a}{2}(1 \pm \sqrt{5}) \end{aligned}$$

Note that $\phi = \frac{1+\sqrt{5}}{2}$ is called the Golden Ratio[5].

Corollary. *2.11.2. The segment $\overline{CF}$ is divided in "extreme and mean ratio" at A.*

---

[5]https://en.wikipedia.org/wiki/Golden_ratio

2.3. PROPOSITIONS FROM BOOK II
131

Corollary. *2.11.3. If from the greater segment $\overline{CA}$ of $\overline{CF}$ we take a segment equal to $\overline{AF}$, it is evident that $\overline{CA}$ will be divided into parts respectively equal to $\overline{AH}$, $\overline{HB}$. Hence, if a segment is divided in extreme and mean ratio, the greater segment will be cut in the same manner by taking on it a part equal to the less, and the less will be similarly divided by taking on it a part equal to the difference, and so on.*

Corollary. *2.11.4. Let $\overline{AB}$ be divided in "extreme and mean ratio" at C. It is evident ([2.11], Cor. 2) that $\overline{AC} > \overline{CB}$. Cut off $\overline{CD} = \overline{CB}$. Then by ([2.11], Cor. 2), $\overline{AC}$ is cut in "extreme and mean ratio" at D, and $\overline{CD} > \overline{AD}$. Next, cut off $\overline{DE} = \overline{AD}$, and in the same manner we have $\overline{DE} > \overline{EC}$, and so on. Since $\overline{CD} > \overline{AD}$, it is evident that $\overline{CD}$ is not a common measure of $\overline{AC}$ and $\overline{CB}$, and therefore not a common measure of $\overline{AB}$ and $\overline{AC}$. Similarly, $\overline{AD}$ is not a common measure of $\overline{AC}$ and $\overline{CD}$ and so is therefore not a common measure of $\overline{AB}$ and $\overline{AC}$. Hence, no matter how far we proceed, we cannot arrive at any remainder which will be a common measure of $\overline{AB}$ and $\overline{AC}$. Hence, the parts of a line divided in "extreme and mean ratio" are incommensurable (i.e., their ratio will never be a rational number).*

Figure 2.3.12: [2.11, Cor. 4]

See also [6.30] where we divide a given segment ($\overline{AB}$) into its "extreme and mean ratio"; that is, we divide $\overline{AB}$ at point C such that $\overline{AB} \cdot \overline{BC} = (\overline{AC})^2$.

Exercises.

1. The difference between the areas of the squares on the segments of a line divided in "extreme and mean ratio" is equal to the area of their rectangle.

2. In a right triangle, if the square on one side is equal in area to the rectangle contained by the hypotenuse and the other side, the hypotenuse is cut in "extreme and mean ratio" by the perpendicular on it from the right angle.

3. If $\overline{AB}$ is cut in "extreme and mean ratio" at H, prove that

   (a) $(\overline{AB})^2 + (\overline{BH})^2 = 3 \cdot (\overline{AH})^2$

   (b) $(\overline{AB} + \overline{BH})^2 = 5 \cdot (\overline{AH})^2$

   [See the final chapter for a solution to (a).]

4. The three lines joining the pairs of points G, B; F, D; A, K, in the construction of [2.11] are parallel.

5. If $\overline{CH}$ intersects $\overline{BE}$ at O, then $\overline{AO} \perp \overline{CH}$.

6. If $\overline{CH}$ is extended, then $\overline{CH} \perp \overline{BF}$.

7. Suppose that $\triangle ABC$ is a right-angled triangle having $\overline{AB} = 2 \cdot \overline{AC}$. If $\overline{AH}$ is equal to the difference between $\overline{BC}$ and $\overline{AC}$, then $\overline{AB}$ is divided in "extreme and mean ratio" at H.

8. Prove [Cor. 2.11.1].

9. Prove [Cor. 2.11.2].

10. Prove [Cor. 2.11.3].

11. Prove [Cor. 2.11.4].

Proposition 2.12. *On an obtuse triangle ($\triangle ABC$), the square on the side opposite the obtuse angle (AB) exceeds the sum of the areas of the squares on the sides containing the obtuse angle (BC, CA) by twice the area of the rectangle contained by either of them (BC) and its extension ($\overline{CD}$) to meet a perpendicular ($\overline{AD}$) on it from the opposite angle.*

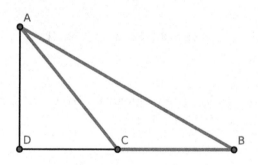

Figure 2.3.13: [2.12]

Corollary. *2.12.1. Algebraically, [2.12] states that in an obtuse triangle*

$$(\overline{AB})^2 = (\overline{AC})^2 + (\overline{BC})^2 + 2 \cdot \overline{BC} \cdot \overline{CD}$$

This is extremely close to stating the law of cosines[6]: $c^2 = a^2 + b^2 - 2ab \cdot cos(\alpha)$

Corollary. *2.12.2. If perpendiculars from A and B to the opposite sides meet them in H and D, the rectangle $AC \cdot CH$ is equal in area to the rectangle $BC \cdot CD$ (or $\square AC \cdot CH = \square BC \cdot CD$).*

---
[6]https://en.wikipedia.org/wiki/Trigonometric_functions

Exercises.

1. If the angle $\angle ACB$ of a triangle is equal to twice the angle of an equilateral triangle, then $AB^2 = BC^2 + CA^2 + BC \cdot CA$.

2. Suppose that $ABCD$ is a quadrilateral whose opposite angles at points B and D are right, and when $\overline{AD}, \overline{BC}$ are extended meet at E, prove that $\overline{AE} \cdot \overline{DE} = \overline{BE} \cdot \overline{CE}$.

3. If $\triangle ABC$ is a right triangle and $\overline{BD}$ is a perpendicular on the hypotenuse AC, prove that $\overline{AB} \cdot \overline{DC} = \overline{BD} \cdot \overline{BC}$.

4. If a segment $\overline{AB}$ is divided at C so that $(\overline{AC})^2 = 2 \cdot (\overline{BC})^2$, prove that $(\overline{AB})^2 + (\overline{BC})^2 = 2 \cdot \overline{AB} \cdot \overline{AC}$.

5. If $\overline{AB}$ is the diameter of a semicircle, find a point C in $\overline{AB}$ such that, joining C to a fixed point D in the circumference and constructing a perpendicular $\overline{CE}$ intersecting the circumference at E, then $(\overline{CE})^2 - (\overline{CD})^2$ is equal to a given square.

6. If the square of a segment $\overline{CD}$, constructed from the angle C of an equilateral triangle $\triangle ABC$ to a point D on the extended side $\overline{AB}$ is equal in area to $2 \cdot (\overline{AB})^2$, prove that $\overline{AD}$ is cut in "extreme and mean ratio" at B.

7. Prove [Cor. 2.12.1].

8. Prove [Cor. 2.12.2].

Proposition 2.13. *In any triangle ($\triangle ABC$), the square on any side opposite an acute angle (at C) is less than the sum of the squares on the sides containing that angle by twice the area of the rectangle ($\overline{BC}$, $\overline{CD}$) contained by either of them (BC) and the intercept ($\overline{CD}$) between the acute angle and the foot of the perpendicular on it from the opposite angle.*

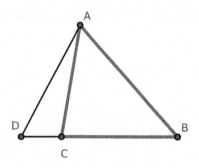

Figure 2.3.14: [2.13]

Corollary. *2.13.1. Algebraically, [2.13] states that in an acute triangle*

$$(\overline{AB})^2 = (\overline{AC})^2 + (\overline{BC})^2 + 2 \cdot \overline{BC} \cdot \overline{CD}$$

which repeats the result of [2.12].

Exercises.

1. If the angle at point C of the $\triangle ACB$ is equal to an angle of an equilateral triangle, then $AB^2 = AC^2 + BC^2 - AC.BC$.

2. The sum of the squares on the diagonals of a quadrilateral, together with four times the square on the line joining their midpoints, is equal to the sum of the squares on its sides.

3. Find a point C in a given extended segment AB such that $AC^2 + BC^2 = 2AC.BC$.

2.3. PROPOSITIONS FROM BOOK II

Proposition 2.14. *CONSTRUCTION OF A SQUARE II. It is possible to construct a square equal in area to any given polygon.*

Proof. We wish to construct a square equal in area to polygon $MNPQ$.

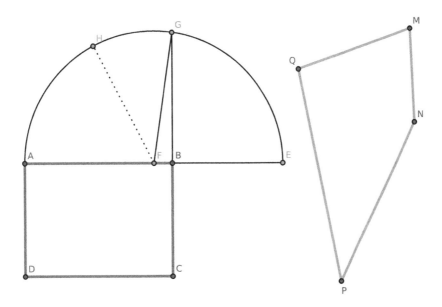

Figure 2.3.15: [2.14]

Construct rectangle $\square ABCD$ equal in area to $MNPQ$ [1.45]. If any two adjacent sides of $\square ABCD$ are equal, then $\square ABCD$ is a square and we have completed the construction.

Otherwise, extend side AB to $\overline{AE}$ such that $\overline{BE} = \overline{BC}$. Bisect $\overline{AE}$ at F, and with F as center and $\overline{FE}$ as radius, construct semicircle AGE. Extend $\overline{CB}$ to the semicircle at G. We claim that the square constructed on $\overline{BG}$ is equal in area to $MNPQ$.

To see this, construct $\overline{FG}$. Because $\overline{AE}$ is divided equally at F and unequally at B, by [2.5] $\overline{AB} \cdot \overline{BE} + (\overline{FB})^2 = (\overline{FE})^2$. Also, $(\overline{FE})^2 = (\overline{FG})^2$, since both are radii of semicircle AGE. By [1.47], $(\overline{FG})^2 = (\overline{FB})^2 + (\overline{BG})^2$. Therefore,

$$\begin{aligned} \overline{AB} \cdot \overline{BE} + (\overline{FB})^2 &= (\overline{FB})^2 + (\overline{BG})^2 \\ \overline{AB} \cdot \overline{BE} &= (\overline{BG})^2 \\ \overline{AB} \cdot \overline{BC} &= (\overline{BG})^2 \\ \square ABCD &= (\overline{BG})^2 \end{aligned}$$

which completes the construction. □

Corollary. *2.14.1. The square on the perpendicular from any point on a semicircle to the diameter is equal to the rectangle contained by the segments of the diameter.*

Exercises.

1. Given the difference of the squares on two segments and their rectangle, find the segments.

2. Prove [Cor. 2.14.1].

Chapter 2 exam questions.

1. What is the subject-matter of chapter 2? (Ans. Theory of rectangles.)

2. What is a rectangle? A gnomon?

3. What is a square inch? A square foot? A square mile? (Ans. The square constructed on a line whose length is an inch, a foot, or a mile.)

4. When is a line said to be divided internally? When externally?

5. How is the area of a rectangle determined?

6. How is a line divided so that the rectangle contained by its segments is a maximum?

7. How is the area of a parallelogram found?

8. What is the altitude of a parallelogram whose base is 65 meters and area 1430 square meters?

9. How is a segment divided when the sum of the squares on its subsegments is a minimum?

10. The area of a rectangle is $108 \cdot 60$ square meters and its perimeter is $48 \cdot 20$ linear meters. Find its dimensions.

11. What proposition in chapter 2 expresses the distributive law of multiplication?

12. On what proposition is the rule for extracting the square root founded?

13. Compare [1.47], [2.12], and [2.13].

14. If the sides of a triangle are expressed algebraically by $x^2 + 1$, $x^2 - 1$, and $2x$ units, respectively, prove that it is a right triangle.

15. How would you construct a square whose area would be exactly an acre? Give a solution using [1.47].

16. What is meant by incommensurable lines? Give an example from chapter 2.

2.3. PROPOSITIONS FROM BOOK II

17. Prove that a side and the diagonal of a square are incommensurable.

18. The diagonals of a lozenge are 16 and 30 meters respectively. Find the length of a side.

19. The diagonal of a rectangle is 4.25 inches, and its area is 7.50 square inches. What are its dimensions?

20. The three sides of a triangle are 8, 11, 15. Prove that it has an obtuse angle.

21. The sides of a triangle are 13, 14, 15. Find the lengths of its medians. Also find the lengths of its perpendiculars and prove that all its angles are acute.

22. If the sides of a triangle are expressed by $m^2 + n^2$, $m^2 - n^2$, and $2mn$ linear units, respectively, prove that it is right-angled.

Chapter 2 exercises.

1. The squares on the diagonals of a quadrilateral are together double the sum of the areas of the squares on the segments joining the midpoints of opposite sides.

2. If the medians of a triangle intersect at O, then $(\overline{AB})^2 + (\overline{BC})^2 + (\overline{CA})^2 = 3((\overline{OA})^2 + (\overline{OB})^2 + (\overline{OC})^2)$.

3. Through a given point O, construct three segments $\overline{OA}$, $\overline{OB}$, $\overline{OC}$ of given lengths such that their endpoints are collinear and that $\overline{AB} = \overline{BC}$.

4. If in any quadrilateral two opposite sides are bisected, the sum of areas of the squares on the other two sides, together with the sum of areas of the squares on the diagonals, is equal to the sum of the areas of the squares on the bisected sides together with four times the area of the square on the line joining the points of bisection.

5. If squares are constructed on the sides of any triangle, the sum of the areas of the squares on the segments joining the adjacent corners is equal to three times the sum of the areas of the squares on the sides of the triangle.

6. Divide a given segment into two parts so that the rectangle contained by the whole and one segment is equal in area to any multiple of the square on the other segment.

7. If P is any point in the diameter $\overline{AB}$ of a semicircle and $\overline{CD}$ is any parallel chord, then $(\overline{CP})^2 + (\overline{PD})^2 = (\overline{AP})^2 + (\overline{PB})^2$.

8. If A, B, C, D are four collinear points taken in order, then $AB \cdot CD + BC \cdot AD = AC \cdot BD$.

9. Three times the sum of the area of the squares on the sides of any pentagon exceeds the sum of the area of the squares on its diagonals by four times the sum of the area of the squares on the segments joining the midpoints of the diagonals.

10. In any triangle, three times the sum of the area of the squares on the sides is equal to four times the sum of the area of the squares on the medians.

11. If perpendiculars are constructed from the vertices of a square to any line, the sum of the squares area of the on the perpendiculars from one pair of opposite angles exceeds twice the area of the rectangle of the perpendiculars from the other pair by the area of the square.

12. If the base AB of a triangle is divided at D such that $m \cdot \overline{AD} = n \cdot \overline{BD}$, then $m \cdot (\overline{AC})^2 + n \cdot (\overline{BC})^2 = m \cdot (\overline{AD})^2 + n \cdot (\overline{DB})^2 + (m+n) \cdot (\overline{CD})^2$.

13. If the point D is taken on the extended segment $\overline{AB}$ such that $m \cdot \overline{AD} = n \cdot \overline{BD}$, then $m \cdot (\overline{AC})^2 - n \cdot (\overline{BC})^2 = m \cdot (\overline{AD})^2 - n \cdot (\overline{DB})^2 + (m-n) \cdot (\overline{CD})^2$.

14. Given the base of a triangle in magnitude and position as well as the sum or the difference of m times the square on one side and n times the square on the other side in magnitude, then the locus of the vertex is a circle.

15. Any rectangle is equal in area to half the rectangle contained by the diagonals of squares constructed on its adjacent sides. [See the final chapter for a solution.]

16. If $A, B, C, ...$ are any finite number of fixed points and P a movable point, find the locus of P if $(\overline{AP})^2 + (\overline{BP})^2 + (\overline{CP})^2 + ...$ is given.

17. If the area of a rectangle is given, its perimeter is a minimum when it is a square.

18. Construct equilateral triangles on subsegments $\overline{AC}, \overline{CB}$ of segment $\overline{AB}$. Prove that if D, D_1 are the centers of circles constructed about these triangles, then $6 \cdot D \cdot D_1^2 = (\overline{AB})^2 + (\overline{AC})^2 + (\overline{CB})^2$.

19. If a, b denote the sides of a right triangle about the right angle and p denotes the perpendicular from the right angle on the hypotenuse, then $\frac{1}{a^2} + \frac{1}{b^2} = \frac{1}{c^2}$.

20. If upon the greater subsegment $\overline{AB}$ of a segment $\overline{AC}$ which is divided in extreme and mean ratio, an equilateral triangle $\triangle ABD$ is constructed and $\overline{CD}$ is joined, then $(\overline{CD})^2 = 2 \cdot (\overline{AB})^2$.

21. If a variable line, whose endpoints rest on the circumferences of two given concentric circles, stands opposite a right angle at any fixed point, then the locus of its midpoint is a circle.

Chapter 3

Circles

Axioms and Mathematical Properties from chapters 1 and 2 will be assumed and not generally cited. This will be a rule that we will apply to subsequent chapters, *mutatis mutandis*.

Remark. Modern geometry no longer uses Euclid's definitions for curves, tangents, etc. However, for our purposes, the definitions are adequate.

3.1 Definitions

1. *Equal circles* are circles with equal radii.[1]

2. A *chord* of a circle is a segment which intersects two points of the circle's circumference. If the chord is extended to a line, then this line is called a *secant*, and each of the parts into which a secant divides the circumference is called an *arc*—the larger is called the *major conjugate arc*, and the smaller is called the *minor conjugate arc*.

3. A segment, ray, or straight line is said to *touch* a circle when it intersects the circumference of a circle at one and only one point. The segment, ray, or straight line is called a *tangent* to the circle, and the point where it touches the circumference is called the *point of intersection*.

---

[1]This is actually a theorem, and not a definition. If two circles have equal radii, they are evidently congruent figures and therefore equal in all aspects. Using this method to prove the theorem, [3.26]-[3.29] follow immediately.

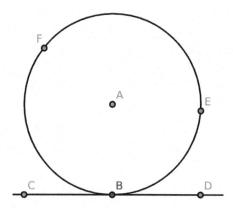

Figure 3.1.1: [Def. 3.3] $\overleftrightarrow{CD}$ touches $\odot A$ at B; or, $\overleftrightarrow{CD}$ is tangent to $\odot A$ and B is the point of intersection between $\odot A$ and $\overleftrightarrow{CD}$.

4. Circles are said to touch one another when they intersect at one and only one point. There are two types of contact:

   a) When one circle is external to the other.

   b) When one circle is internal to the other.

When circles intersect at two points, the intersection may be referred to as a *cut*.

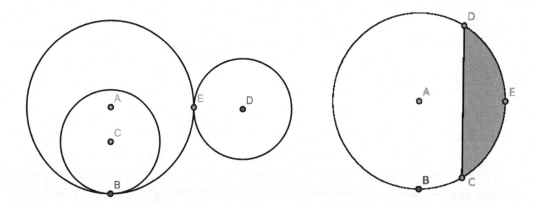

Figure 3.1.2: **On the left:** [Def. 3.4] The circles $\odot A$ and $\odot D$ touch externally at E, while the circles $\odot C$ and $\odot A$ touch internally at B. **On the right:** [Def. 3.5] The chord $\overline{CD}$ of the circle $\odot A$ divides the circle itself into segments DEC and DBC. Segment DEC (shaded) is bounded by chord $\overline{CD}$ and arc DEC, and segment DBC (unshaded) is bounded by chord $\overline{CD}$ and arc DBC.

5. A *segment of a circle* is a two-dimensional figure bounded by a chord and an arc whose boundary points include the endpoints of the chord.

6. Chords are said to be equally distant from the center when the perpendiculars constructed to them from the center are equal in length.

3.1. DEFINITIONS

7. The *angle in the segment* is the rectilinear angle contained between two chords which intersect at the same endpoint on the circumference of a circle. In Fig. 3.1.3, $\angle DCE$ is an angle in the segment. See also [3.21].

8. The *angle of a segment* is the non-rectilinear angle contained between its chord and the tangent at either endpoint. In Fig. 3.1.3, the arc DEC is the angle of segment DEC. These angles only appeared in the original proof to [3.16].

9. An angle in a segment is said to *stand* on its conjugate arc.

10. Similar arcs are those that contain equal angles.

11. A sector of a circle is formed by two radii and the arc that is included between them. In Fig. 3.1.3, $\odot A$, radius $\overline{AD}$, and radius $\overline{AC}$ form the sectors $DACE$ and $DACB$.

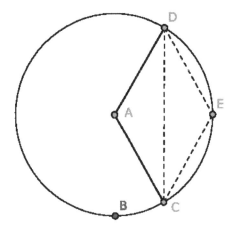

Figure 3.1.3: [Def. 3.8], [Def. 3.9], and [Def. 3.11]

12. *Concentric circles* are those which have the same center point.

13. Points which lie on the circumference of a circle are called *concyclic*.

14. A *cyclic quadrilateral* is a quadrilateral which is inscribed in a circle.

15. A modern definition on an angle[2]: an *angle* in geometry is the figure formed by two rays, called the sides of the angle, which share a common endpoint, called the vertex of the angle. This measure is the ratio of the length of a circular arc to its radius, where the arc is centered at the vertex and delimited by the sides.

The size of a geometric angle is usually characterized by the magnitude of the smallest rotation that maps one of the rays into the other. Angles that have the same size are called *congruent angles*.

---
[2]http://en.wikipedia.org/wiki/Angle

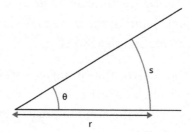

Figure 3.1.4: The measure of angle θ is the quotient of s and r.

In order to measure an angle θ, a circular arc centered at the vertex of the angle is constructed, e.g., with a pair of compasses. The length of the arc is then divided by the radius of the arc r, and possibly multiplied by a scaling constant k (which depends on the units of measurement that are chosen):

$$\theta = ks/r$$

The value of θ thus defined is independent of the size of the circle: if the length of the radius is changed, then the arc length changes in the same proportion, and so the ratio s/r is unaltered.

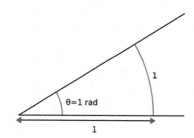

Figure 3.1.5: $\theta = s/r$ rad $= 1$ rad.

A number of units are used to represent angles: the radian and the degree are by far the most commonly used.

Most units of angular measurement are defined such that one turn (i.e. one full circle) is equal to n units, for some whole number n. In the case of degrees, $n = 360$. A turn of n units is obtained by setting $k = \frac{n}{2\pi}$ in the formula above.

The radian is the angle subtended by an arc of a circle (that is, the angle standing opposite the arc of a circle) that has the same length as the circle's radius. The case of radian for the formula given earlier, a radian of $n = 2\pi$ units is obtained by setting $k = \frac{2\pi}{2\pi} = 1$. One turn is 2π radians, and one radian is $\frac{180}{\pi}$ degrees, or about 57.2958 degrees. The radian is abbreviated rad, though this symbol is often omitted in mathematical texts, where radians are assumed unless specified otherwise. When

3.1. DEFINITIONS

radians are used angles are considered as dimensionless. The radian is used in virtually all mathematical work beyond simple practical geometry, due, for example, to the pleasing and "natural" properties that the trigonometric functions display when their arguments are in radians. The radian is the (derived) unit of angular measurement in the SI system.

The degree, written as a small superscript circle (°), is 1/360 of a turn, so one turn is 360°. Fractions of a degree may be written in normal decimal notation (e.g. 3.5° for three and a half degrees), but the "minute" and "second" sexagesimal sub-units of the "degree-minute-second" system are also in use, especially for geographical coordinates and in astronomy and ballistics.

Although the definition of the measurement of an angle does not support the concept of a negative angle, it is frequently useful to impose a convention that allows positive and negative angular values to represent orientations and/or rotations in opposite directions relative to some reference.

In a two-dimensional Cartesian coordinate system, an angle is typically defined by its two sides, with its vertex at the origin. The initial side is on the positive x-axis, while the other side or terminal side is defined by the measure from the initial side in radians, degrees, or turns. Positive angles represent rotations toward the positive y-axis, and negative angles represent rotations toward the negative y-axis. When Cartesian coordinates are represented by standard position, defined by the x-axis rightward and the y-axis upward, positive rotations are anticlockwise and negative rotations are clockwise.

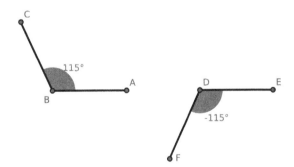

Figure 3.1.6: $\angle CBA$ measured as a positive angle, $\angle EDF$ measured as a negative angle

16. Suppose we have two points F and P such that when the area of the rectangle $OF \cdot OP$ is equal to the area of the square of the radius of that circle, then F and P are called *inverse points* with respect to the circle.

17. The supplement of an arc is the amount by which an arc is less than a semicircle, or an angle less than two right angles.

3.2 Propositions from Book III

Proposition 3.1. *THE CENTER OF A CIRCLE I.*

It is possible to locate the center of a circle.

Proof. Construct a circle and take any two points A, B on the circumference. Construct $\overline{AB}$ and bisect $\overline{AB}$ at C [1.10]. Construct $\overline{CD} \perp \overline{AB}$ (where D is on the circumference) and extend $\overline{CD}$ to intersect the circumference at E. Bisect $\overline{DE}$ at F. We claim that F is the center of the circle.

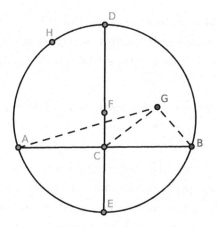

Figure 3.2.1: [3.1]

Suppose instead that point G, which does not lie on chord $\overline{DE}$, is the center of the circle. Construct $\overline{GA}$, $\overline{GC}$, and $\overline{GB}$. Notice that $\angle ACG = \angle ACD + \angle DCG$; clearly, $\angle DCG > 0$.

Consider $\triangle ACG$ and $\triangle BCG$: we have $AC = CB$ by construction, $GA = GB$ (since they are radii by hypothesis), and side CG is shared in common. By [1.8], we find that $\angle ACG = \angle BCG$; therefore, each angle is a right angle. But $\angle ACD$ is right by construction; therefore $\angle DCG = 0$. But $\angle DCG > 0$ above, a contradiction.

Hence, no point can be the center of the circle other than a point on $\overline{DE}$. Since all radii are equal in length and $\overline{FE} = \overline{FD}$, it follows that F, the midpoint of $\overline{DE}$, is the center of $\bigcirc F$. This proves our claim. □

3.2. PROPOSITIONS FROM BOOK III

Alternatively:

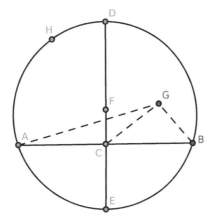

Proof. Consider the objects constructed in the proof above: because $\overline{ED} \perp \overline{AB}$ and $\overline{ED}$ bisects $\overline{AB}$, every point equally distant from the points A and B must lie on $\overline{ED}$ [1.10, #2]. Since the center is equally distant from A and B, the center must lie on $\overline{ED}$. And since the center must also lie equally distant from E and D, the center is the midpoint of $\overline{ED}$. □

Corollary. *3.1.1. The line, ray, or segment which bisects any chord of a circle perpendicularly passes through the center of the circle.*

Corollary. *3.1.2. The locus of the centers of the circles which pass through two fixed points is the line bisecting at right angles the line that connects the two points.*

Corollary. *3.1.3. If A, B, C are three points on the circumference of a circle, the lines which perpendicularly bisect the chords $\overline{AB}$, $\overline{BC}$ will intersect at the center of the circle.*

Exercises.

1. Prove [Cor. 3.1.1].

2. Prove [Cor. 3.1.2].

3. Prove [Cor. 3.1.3].

Proposition 3.2. *POINTS ON A LINE INSIDE AND OUTSIDE A CIRCLE.*

If any two points are chosen from the circumference of a circle and a line is constructed on those points, then:

(1) The points between the endpoints on the circumference form a chord (i.e., they lie inside the circle).

(2) The remaining points of the line lie outside the circle.

Proof. Construct $\odot C$ where A and B are arbitrary points on the circumference of $\odot C$ and construct $\overleftrightarrow{AB}$. We claim that:

(1) $\overline{AB}$ is a chord of $\odot C$.

(2) All points of $\overleftrightarrow{AB}$ which are not on $\overline{AB}$ lie outside of the circle.

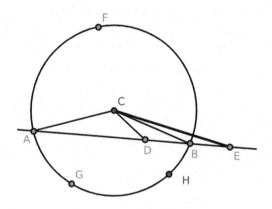

Figure 3.2.2: [3.2]

Take any point D on $\overline{AB}$ and construct $\overline{CA}$, $\overline{CD}$, and $\overline{CB}$. Notice that $\angle ADC > \angle ABC$ by [1.16]; however, $\angle ABC = \angle BAC$ because $\triangle CAB$ is isosceles [1.5]. Therefore, $\angle ADC > \angle BAC = \angle DAC$. By [1.29], $\overline{AC} > \overline{CD}$ [1.29], and so $\overline{CD}$ is less than the radius of $\odot C$. Consequently, D must lie within the circle [Def. 1.23]. Similarly, every other point between A and B lies within $\odot C$. Finally, since A and B are points in the circumference of $\odot C$, $\overline{AB}$ is a chord. This proves claim 1.

Wlog, let E be any point on $\overrightarrow{AE}$ such that $\overline{EA} > \overline{BA}$, and construct $\overline{CE}$. By [1.16], $\angle ABC > \angle AEC$; by the above, $\angle CAE > \angle AEC$. It follows that in $\triangle ACE$, $\overline{CE} > \overline{CA}$, and so the point E lies outside $\odot C$. This proves claim 2, which completes the proof. $\square$

Corollary. *3.2.1. Three collinear points cannot be concyclic.*

Corollary. *3.2.2. A straight line, ray, or segment cannot intersect a circle at more than two points.*

Corollary. *3.2.3. The circumference of a circle is everywhere concave towards the center.*

Exercises.

1. Prove [Cor. 3.1.1].

2. Prove [Cor. 3.1.2].

3. Prove [Cor. 3.1.3].

Proposition 3.3. *CHORDS I.*

Suppose there exist two chords of a circle, only one of which passes through the center of the circle. The chord which does not pass through the center is bisected by the chord that passes through the center if and only if the chords are perpendiculars.

Proof. Construct $\odot O$ with chords $\overline{AB}$ and $\overline{CD}$ where $\overline{AB}$ contains the center of $\odot O$. We claim that $\overline{AB}$ bisects $\overline{CD}$ if and only if $\overline{AB} \perp \overline{CD}$.

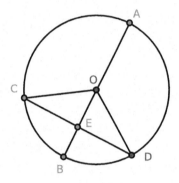

Figure 3.2.3: [3.3]

Suppose that $\overline{AB}$ bisects $\overline{CD}$. Construct $\overline{OC}$ and $\overline{OD}$, and consider $\triangle CEO$ and $\triangle DEO$: $\overline{CE} = \overline{ED}$ by hypothesis, $\overline{OC} = \overline{OD}$ since each are radii of $\odot O$, and both triangles have $\overline{EO}$ in common. By [1.8], $\angle CEO = \angle DEO$; since they are also adjacent angles, each is a right angle, and therefore $\overline{AB} \perp \overline{CD}$.

Now suppose that $\overline{AB} \perp \overline{CD}$. Because $\overline{OC} = \overline{OD}$, $\triangle CDO$ is isosceles; by [1.5], $\angle OCD = \angle ODC$. Consider $\triangle OEC$ and $\triangle OED$: $\angle OCD = \angle ODC$, $\angle CEO = \angle DEO$ since $\overline{AB} \perp \overline{CD}$, and they share side $\overline{EO}$. By [1.26], $\triangle OEC \cong \triangle OED$, and so $\overline{CE} = \overline{ED}$. Since $\overline{CD} = \overline{CE} \oplus \overline{ED}$, $\overline{AB}$ bisects $\overline{CD}$, proving our claim. $\square$

The second part of the proposition may also be proved this way:

Proof. By [1.47], we find that

$$\begin{aligned}(\overline{OC})^2 &= (\overline{OE})^2 + (\overline{EC})^2 \\ (\overline{OD})^2 &= (\overline{OE})^2 + (\overline{ED})^2\end{aligned}$$

Since $\overline{OC} = \overline{OD}$, we also have that $(\overline{OC})^2 = (\overline{OD})^2$, and it follows that $(\overline{EC})^2 = (\overline{ED})^2$. Therefore, $\overline{EC} = \overline{ED}$. $\square$

3.2. PROPOSITIONS FROM BOOK III

Corollary. *3.3.1. The line which bisects perpendicularly one of two parallel chords of a circle bisects the other perpendicularly.*

Corollary. *3.3.2. The locus of the midpoints of a system of parallel chords of a circle is the diameter of the circle perpendicular to them all.*

Corollary. *3.3.3. If a line intersects two concentric circles, its intercepts between the circles are equal in length.*

Corollary. *3.3.4. The line connecting the centers of two intersecting circles bisects their common chord perpendicularly.*

Observation: [3.1], [3.3], and [3.3, Cor. 1] are related such that if any one of them is proved directly, then the other two follow by the Rule of Symmetry.

Exercises.

1. If a chord of a circle stands opposite a right angle at a given point, the locus of its midpoint is a circle.

2. Prove [3.3, Cor. 1].

3. Prove [3.3, Cor. 2].

4. Prove [3.3, Cor. 3].

5. Prove [3.3, Cor. 4]. [See the final chapter for a solution.]

Proposition 3.4. *CHORDS II.*

If two chords exist in a circle where at most one is a diameter, then it is not the case that each chord bisects the other.

Proof. Construct $\odot O$ with chords $\overline{AB}$ and $\overline{CD}$ such that at most one of these chords is a diameter and such that $\overline{AB}$ and $\overline{CD}$ intersect at E where $\angle AEC > 0$. If $\overline{AB}$ and $\overline{CD}$ are not diameters, they do not contain O. Construct $\overline{OE}$ and extend $\overline{OE}$ to $\overline{FG}$. We claim it is not the case that $\overline{AE} = \overline{EB}$ and $\overline{CE} = \overline{ED}$.

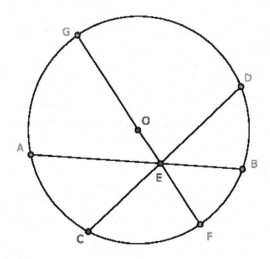

Figure 3.2.4: [3.4]

Suppose instead that $\overline{AE} = \overline{EB}$ and $\overline{CE} = \overline{ED}$. By [3.3], $\angle OEA$ is a right angle. Similarly, $\angle OEC$ is a right angle, or $\angle OEA = \angle OEC$. But $\angle OEC = \angle OEA + \angle AEC$, and so $\angle AEC = 0$. It follows that $\angle AEC > 0$ and $\angle AEC = 0$, a contradiction.

Thus it is not the case that $\overline{AE} = \overline{EB}$ and $\overline{CE} = \overline{ED}$, which completes the proof. $\square$

Corollary. *3.4.1. If two chords of a circle bisect each other, they are both diameters.*

3.2. PROPOSITIONS FROM BOOK III

Proposition 3.5. *NON-CONCENTRIC CIRCLES I.*

If two circles intersect at exactly two points, then they are not concentric.

Proof. Construct $\circ ABC$ and $\circ ABD$ which intersect at A and B; we claim that $\circ ABC$ and $\circ ABD$ are not concentric.

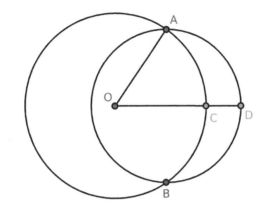

Figure 3.2.5: [3.5]

Suppose instead that $\circ ABC$ and $\circ ABD$ share a common center, O. Construct $\overline{OA}$ and $\overline{OCD}$ where A, B, C, and D are distinct. Notice that $\overline{CD} > 0$.

Because O is the center of $\circ ABC$, $\overline{OA} = \overline{OC}$. Because O is the center of the circle $\circ ABD$, $\overline{OA} = \overline{OD}$; hence, $\overline{OD} = \overline{OC}$ where $\overline{OD} = \overline{OC} \oplus \overline{CD}$. It follows that $\overline{CD} = 0$ and $\overline{CD} > 0$, a contradiction. Therefore, $\circ ABC$ and $\circ ABD$ are not concentric. □

Exercises.

1. Two circles cannot have three points in common without coinciding. [See the final chapter for a solution.]

Proposition 3.6. *NON-CONCENTRIC CIRCLES II.*

If one circle intersects another circle internally at exactly one point, then the circles are not concentric.

Proof. Construct $\circ ADE$ and $\circ ABC$ such that $\circ ABC$ intersects $\circ ADE$ internally at A and only at A. We claim that $\circ ADE$ and $\circ ABC$ are not concentric.

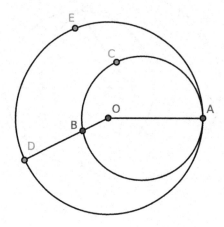

Figure 3.2.6: [3.6]

Suppose instead that $\circ ADE$ and $\circ ABC$ are concentric, and let O be the center of each circle. Construct $\overline{OA}$ and $\overline{OBD}$. Notice that $\overline{BD} > 0$; if $\overline{BD} = 0$, then $B = D$ and $\circ ADE$ and $\circ ABC$ intersect at two points, contrary to hypothesis.

Because O is the center of each circle by hypothesis, $\overline{OA} = \overline{OB}$ and $\overline{OA} = \overline{OD}$; therefore, $\overline{OB} = \overline{OD}$ and $\overline{OB} \oplus \overline{BD} = \overline{OD}$. Hence, $\overline{BD} = 0$, a contradiction. Therefore, the circles are not concentric. □

3.2. PROPOSITIONS FROM BOOK III

Proposition 3.7. *UNIQUENESS OF SEGMENT LENGTHS FROM A POINT ON THE DIAMETER OTHER THAN THE CENTER.*

Choose any point on the diameter of a circle other than the center, and from that point construct a finite number of segments to the circumference. Then:

(1) The longest segment constructed will contain the center of the circle.

(2) The shortest segment constructed will form a diameter with the longest segment.

(3) As for the remaining segments, those with endpoints on the circumference nearer to the endpoint of the longest segment will be longer than segments with endpoints farther from the endpoint of the longest segment.

(4) Two and only two equal segments can be constructed from each point to the circumference, one on each side of the diameter.

Proof. Construct $\odot O$ with point P on diameter $\overline{AE}$ such that O and P are distinct. Construct a finite number of segments from P to the circumference ($\overline{PA}, \overline{PB}, \overline{PC}$, etc.). Notice that $\overline{PA}$ is a segment on the diameter. We will prove four claims.

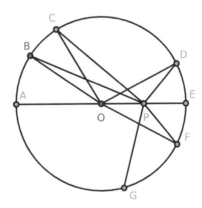

Figure 3.2.7: [3.7] $\odot EAG$

1. The longest segment, $\overline{PA}$, is the segment which passes through O.

Construct $\overline{OB}$ where B is a point on $\odot O$. Clearly, $\overline{OA} = \overline{OB}$. From this we obtain $\overline{PA} = \overline{OA} + \overline{OP} = \overline{OB} + \overline{OP}$. Consider $\triangle OPB$: since $\overline{OB} + \overline{OP} > \overline{PB}$ by [1.20], it follows that $\overline{PA} > \overline{PB}$. Since this inequality holds for any segment constructed using this method, $\overline{PA}$ is the longest segment of all such constructed segments.

2. The extension of $\overline{PA}$ in the opposite direction, $\overline{PE}$, is the shortest segment of all such constructed segments.

Construct $\overline{OD}$ and consider $\triangle OPD$: by [1.20], $\overline{OP} + \overline{PD} > \overline{OD}$. Since, $\overline{OD} = \overline{OE} = \overline{OP} + \overline{PE}$, it follows that

$$\overline{OP} + \overline{PD} > \overline{OP} + \overline{PE}$$
$$\overline{PD} > \overline{PE}$$

Since this inequality holds for any segment constructed using this method, $\overline{PE}$ is the shortest segment of all such constructed segments.

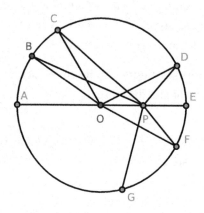

Figure 3.2.8: [3.7] $\circ EAG$

3. As for the remaining segments, those with endpoints on the circumference nearer to the endpoint of the longest segment ($\overline{PA}$) will be longer than segments with endpoints farther from the endpoint of the longest segment (i.e., $\overline{PA} > \overline{PB} > \overline{PC} > \overline{PD}$).

Construct $\overline{OC}$, and consider $\triangle POB$ and $\triangle POC$: $\overline{OB} = \overline{OC}$ and each shares side $\overline{OP}$. Since $\angle POB = \angle POC + \angle BOC$, we find that $\angle POB > \angle POC$. By [1.24], $\overline{PB} > \overline{PC}$. Similarly, $\overline{PC} > \overline{PD}$.

4. Two and only two segments making equal angles with the diameter and standing on opposite sides of the diameter are equal in length (i.e., $\overline{PD} = \overline{PF}$).

At O, construct $\angle POF = \angle POD$ and construct $\overline{PF}$. Consider $\triangle POD$ and $\triangle POF$: $\overline{OD} = \overline{OF}$, each shares side $\overline{OP}$, and $\angle POD = \angle POF$ by construction. By [1.4], $\triangle POD \cong \triangle POF$, and so $\angle OPF = \angle OPD$ and $\overline{PD} = \overline{PF}$.

We claim that a third segment cannot be constructed from P equal to $\overline{PD} = \overline{PF}$. Suppose this were possible and let $\overline{PG} = \overline{PD}$. Then $\overline{PG} = \overline{PF}$, contradicting claim 3 above.

This completes the proof. □

3.2. PROPOSITIONS FROM BOOK III

Corollary. *3.7.1. If two equal segments $\overline{PD}$, $\overline{PF}$ are constructed from a point P to the circumference of a circle, the diameter through P bisects the angle $\angle DPF$ formed by these segments.*

Corollary. *3.7.2. If P is the common center of circles whose radii are $\overline{PA}$, $\overline{PB}$, $\overline{PC}$, $\overline{PD}$, etc., then:*

(1) The circle whose radius is the maximum segment ($\bigcirc P$ with radius $\overline{PA}$) lies outside $\bigcirc O$ and intersects it at A [Def. 3.4].

(2) The circle whose radius is the minimum segment ($\bigcirc P$ with radius $\overline{PE}$) lies inside $\bigcirc O$ and intersects it at E.

(3) A circle having any of the remaining radii (such as $\overline{PD}$) cuts $\bigcirc O$ at two points (such as D, F).

Exercises.

1. Prove [Cor. 3.7.1].

2. Prove [Cor. 3.7.2].

Remark. [3.7] is a good illustration of the following important definition: if a geometrical magnitude varies its position continuously according to any well-defined relationship, and if it retains the same value throughout, it is said to be a constant (such as the radius of a fixed circle).

But if a magnitude increases for some time and then begins to decrease, it is said to be a maximum when the increase stops. Therefore in the previous figure, $\overline{PA}$, which we suppose to revolve around P and meet the circle, is a maximum.

Again, if it decreases for some time, and then begins to increase, it is a minimum at the beginning of the increase. Thus $\overline{PE}$, which we suppose as before to revolve around P and meet the circle, is a minimum. [3.8] will provide other illustrations of this concept.

Proposition 3.8. *SEGMENT LENGTHS FROM A POINT OUTSIDE THE CIRCLE AND THEIR UNIQUENESS.*

Suppose a point is chosen outside of a circle and from that point segments are constructed such that they intersect the circumference of the circle at two points, one on the "outer" or convex side of the circumference and one on the "inner" or concave side of the circumference. Let one segment be constructed which intersects the center of the circle and the others all within the same semicircle but not through the center of the circle. Then:

(1) The largest segment passes through the center.

(2) The segments nearer to the segment through the center are greater in length than those which are farther away.

(3) If segments are constructed to the convex circumference, the minimum segment is that which passes through the center when extended.

(4) Of the other segments, that which is nearer to the minimum is smaller than one more farther out.

(5) From the given point outside of the circle, there can be constructed two equal segments to the concave or the convex circumference, both of which make equal angles with the line passing through the center.

(6) Three or more equal segments cannot be constructed from the given point outside the circle to either circumference.

Proof. Construct $\odot O$, point P outside of $\odot O$, and all points indicated in the figure below. We will prove each claim separately.

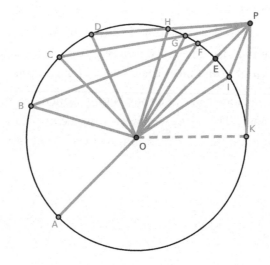

Figure 3.2.9: [3.8] $\odot O$

3.2. PROPOSITIONS FROM BOOK III

1. The maximum segment passes through the center.

Notice that $\overline{OA} = \overline{OB}$ and so $\overline{AP} = \overline{OA} + \overline{OP} = \overline{OB} + \overline{OP}$. Consider $\triangle BOP$: $\overline{OB} + \overline{OP} > \overline{BP}$ [1.20]. Therefore, $\overline{AP} > \overline{BP}$.

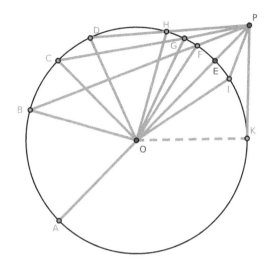

Figure 3.2.10: [3.8] $\odot O$

2. The segments nearer to the segment through the center are greater in length than those which are farther.

Consider $\triangle BOP$ and $\triangle COP$: $\overline{OB} = \overline{OC}$, each share side $\overline{OP}$, and the angle $\angle BOP > \angle COP$. Therefore, $\overline{BP} > \overline{CP}$ [1.24]. Similarly, $\overline{CP} > \overline{DP}$, etc.

3. If segments are constructed to the convex circumference, the minimum segment is that which passes through the center when extended.

Consider $\triangle OFP$: $\overline{OF} + \overline{FP} > \overline{OP} = \overline{OE} + \overline{EP}$ [1.20]. Since $\overline{OF} = \overline{OE}$, we find that $\overline{FP} > \overline{EP}$.

4. Of the other segments, that which is nearer to the minimum is smaller than one more farther out.

Consider $\triangle GOP$ and $\triangle FOP$: $\overline{GO} = \overline{FO}$, each shares side $\overline{OP}$, and the angle $\angle GOP > \angle FOP$. By [1.24], $\overline{GP} > \overline{FP}$. Similarly, $\overline{HP} > \overline{GP}$.

5. From the given point outside of the circle, there can be constructed two equal segments to the concave or the convex circumference, both of which make equal angles with the line passing through the center.

Construct $\angle POI$ such that $\angle POI = \angle POF$ [1.23], and consider $\triangle IOP$ and $\triangle FOP$: $\overline{IO} = \overline{FO}$, each shares side $\overline{OP}$, and $\angle IOP = \angle POF$ by construction. By [1.4], $\overline{IP} = \overline{FP}$. Segments $\overline{IP}$ and $\overline{FP}$ fulfill the above requirements, which proves this claim.

6. Three or more equal segments cannot be constructed from the given point outside the circle to either circumference.

Above, we obtain $\overline{IP} = \overline{FP}$. We claim that a third segment cannot is constructed from P equal to $\overline{IP}$ and $\overline{FP}$. Suppose this were possible and let $\overline{PK} = \overline{PF}$. However by claim 4, $\overline{PK} > \overline{PF}$. This contradiction proves our claim. □

Corollary. *3.8.1. If $\overline{PI}$ is extended to meet the circle at L, then $\overline{PL} = \overline{PB}$.*

Corollary. *3.8.2. If two equal segments are constructed from P to either the convex or concave circumference, the diameter through P bisects the angle between them, and the segments intercepted by the circle are equal in length.*

Corollary. *3.8.3. If P is the common center of circles whose radii are segments constructed from P to the circumference of $\odot O$, then:*

a) The circle whose radius is the minimum segment ($\overline{PE}$) has external contact with $\odot O$ [Def. 3.4].

b) The circle whose radius is the maximum segment ($\overline{PA}$) has internal contact with $\odot O$.

c) A circle having any of the remaining segments ($\overline{PF}$) as radius intersects $\odot O$ at two points (F, I).

Exercises.

1. Prove [Cor. 3.8.1].

2. Prove [Cor. 3.8.2].

3. Prove [Cor. 3.8.3].

3.2. PROPOSITIONS FROM BOOK III

Proposition 3.9. *THE CENTER OF A CIRCLE II.*

A point within a circle from which three or more equal segments can be constructed to the circumference is the center of that circle.

Proof. Construct $\bigcirc ABC$ and equal segments $\overline{DA}$, $\overline{DB}$, and $\overline{DC}$. We claim that D is the center of $\bigcirc ABC$.

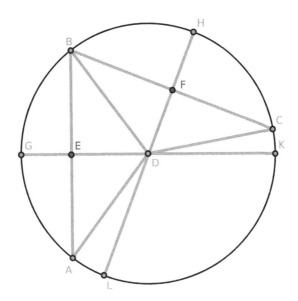

Figure 3.2.11: [3.9]

Construct $\overline{AB}$ and $\overline{BC}$ and bisect them at points E and F, respectively [1.10]. Then construct $\overline{GEDK}$ and $\overline{LDFH}$.

Consider $\triangle AED$ and $\triangle BED$: $\overline{ED}$ is a common side, $\overline{AE} = \overline{EB}$, and $\overline{DB} = \overline{DA}$ since each are radii of $\bigcirc ABC$. By [1.8], $\triangle AED \cong \triangle BED$, and so $\angle AED = \angle BED$; it follows that $\angle AED$ and $\angle BED$ are each right angles.

Since $\overline{GEDK} \perp \overline{AB}$ and $\overline{GEDK}$ bisects $\overline{AB}$, [3.1, Cor. 1] states that the center of $\bigcirc ABC$ is a point on $\overline{GEDK}$. Similarly, the center of $\bigcirc ABC$ is a point on $\overline{LDFH}$. Since $\overline{GEDK}$ and $\overline{LDFH}$ intersect at D, D is the center of $\bigcirc ABC$. $\square$

Alternatively:

Proof. Since $\overline{AD} = \overline{LD}$, the segment bisecting the angle $\angle ADL$ passes through the center [3.7, Cor. 1]. Similarly, the segment bisecting the angle $\angle BDA$ passes through the center. Hence, the point of intersection of these bisectors, D, is the center. $\square$

Proposition 3.10. *THE UNIQUENESS OF CIRCLES.*

If two circles have more than two points of their circumferences in common, they coincide.

Proof. Construct $\bigcirc ABC$ and $\bigcirc DAB$ such that they have more than two points in common. We claim that $\bigcirc ABC$ and $\bigcirc DAB$ coincide.

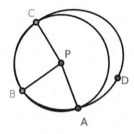

Figure 3.2.12: [3.10]

Suppose that $\bigcirc ABC$ and $\bigcirc DAB$ share three points in common (A, B, C). From P, the center of $\bigcirc ABC$, construct $\overline{PA}$, $\overline{PB}$, $\overline{PC}$; since each is a radius, $\overline{PA} = \overline{PB} = \overline{PC}$.

Since $\bigcirc DAB$ is a circle and P a point from which three equal segments $\overline{PA}$, $\overline{PB}$, $\overline{PC}$ can be constructed to its circumference, P is also the center of $\bigcirc DAB$ [3.9]. By [Def. 3.1] $\bigcirc ABC$ and $\bigcirc DAB$ coincide, which proves our claim. □

Corollary. *3.10.1. Two circles which do not coincide do not have more than two points common.*

Remark. Similarly to [3.10, Cor. 1], two lines which do not coincide cannot have more than one point common.

Exercise.

1. Prove [Cor. 3.10.1].

3.2. PROPOSITIONS FROM BOOK III   161

Proposition 3.11. *SEGMENTS CONTAINING CENTERS OF CIRCLES.*

If one circle touches another circle internally at one point, then the line joining the centers of the two circles must contain that point of intersection.

Proof. Construct $\odot O$ and $\odot H$ such that $\odot H$ touches $\odot O$ internally at P. Also construct $\overleftrightarrow{OH}$. We claim that $\overleftrightarrow{OH}$ contains P.

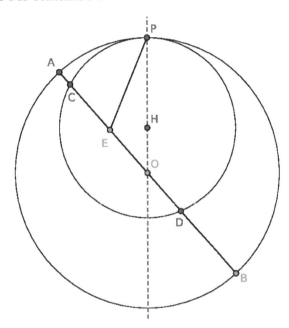

Figure 3.2.13: [3.11]

Suppose instead that E is the center of $\odot H$ such that E does not lie on $\overleftrightarrow{OP}$, and construct $\overline{EP}$. Extend $\overline{OE}$ to intersect $\odot H$ at C and D and intersect $\odot O$ at A and B. Since E is a point on the diameter of $\odot O$ between O and A, $\overline{EA} < \overline{EP}$ [3.7].

Notice that $\overline{EA} = \overline{EC} + \overline{CA}$. Since $\overline{CA} > 0$, $\overline{EA} > \overline{EC}$. Also notice that $\overline{EP} = \overline{EC}$ since each are radii of $\odot H$, and so $\overline{EA} > \overline{EP}$. But $\overline{EA} < \overline{EP}$ above; this contradiction demonstrates that the center of the internal circle, H, must lie on $\overleftrightarrow{OP}$; that is, $\overleftrightarrow{OP}$ contains H. Equivalently, $\overleftrightarrow{OH}$ contains P, which concludes the proof.   □

Alternatively:

Proof. Since $\overline{EP}$ is a segment constructed from a point within the circle $\odot O$ to the circumference of $\odot O$ but not forming part of the diameter through E, the circle whose center is E with radius $\overline{EP}$ cuts $\odot H$ at P [3.7, Cor. 2] and also touches it at P by hypothesis, a contradiction. A similar argument holds for all points not on $\overline{OP}$. Hence, the center of $\odot O$ must lie on $\overline{OP}$.   □

162   CHAPTER 3. CIRCLES

Proposition 3.12. *INTERSECTING CIRCLES I.*

If two circles intersect externally at one point, then the segment joining their centers contains that point of intersection.

Proof. Construct $\odot PCF$ and $\odot PDE$ which intersect externally at point P. Also construct $\overline{AG}$. We claim that $\overline{AG}$ contains P.

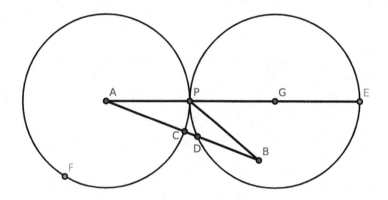

Figure 3.2.14: [3.12]

Construct $\overline{APE}$; a claim equivalent to the above is that $\overline{APE}$ contains G.

Suppose instead that B is the center of $\odot PDE$ and is not on $\overline{APE}$. Construct $\overline{AB}$, intersecting $\odot PDE$ at D and $\odot PCF$ at C but not intersecting at P. Also construct $\overline{BP}$. By our hypothesis, $\overline{AP} = \overline{AC}$ and $\overline{BP} = \overline{BD}$. Hence

$$\overline{AC} + \overline{DB} = \overline{AP} + \overline{BP}$$

Also notice that $\overline{AB} = \overline{AC} \oplus \overline{CD} \oplus \overline{DB}$ where $\overline{CD} > 0$; it follows that $\overline{AB} > \overline{AC} + \overline{DB}$. By the above equality, $\overline{AB} > \overline{AP} + \overline{BP}$.

Consider $\triangle APB$: we find that one side of $\triangle APB$ is greater than the sum of the other two, contradicting [1.20]. Thus, the center of $\odot PDE$ lies on $\overline{APE}$ at G. Equivalently, $\overline{AG}$ contains P, which completes the proof. $\square$

Alternatively:

Proof. Suppose that the center of $\odot PDE$ lies on $\overline{BP}$. Since $\overline{BP}$ is a segment constructed from a point outside of the circle $\odot PCF$ to its circumference which does not pass through the center when it is extended, the circle whose center is B with radius $\overline{BP}$ must cut the circle $\odot PCF$ at P [3.8, Cor. 3].

However, such a circle touches $\odot PCF$ at P by hypothesis, a contradiction. Since $\overline{BP}$ was chosen as any segment other than $\overline{PE}$, the center of $\odot PDE$ must lie on $\overline{PE}$. $\square$

3.2. PROPOSITIONS FROM BOOK III   163

Remark. [3.11] and [3.12] may written as one theorem: "If two circles touch each other at any point, the centers and that point are collinear." This is a limiting case of the theorem given in [3.3, Cor. 4]: "The line joining the centers of two intersecting circles bisects the common chord perpendicularly."

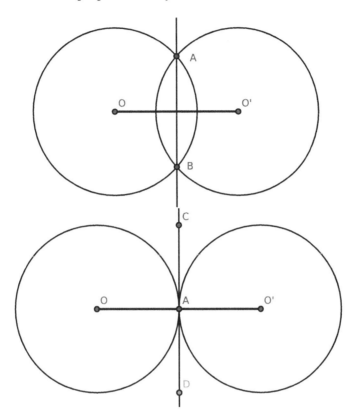

Figure 3.2.15: [3.12], Suppose $\odot O$ and $\odot O'$ have two points of intersection, A and B. Suppose further that A remains fixed while the second circle moves so that the point B ultimately coincides with A. Since the segment $\overline{OO'}$ always bisects $\overline{AB}$, we see that $\overline{OO'}$ intersects A. In consequence of this motion, the common chord $\overline{CD}$ becomes the tangent to each circle at A.

Corollary. *3.12.1. If two circles touch each other, their point of intersection is the union of two points of intersection. When counting the number of points at which two circles intersect, we may for purposes of calculation consider this point of intersection as two points.*

Corollary. *3.12.2. If two circles touch each other at any point, they cannot have any other common point.*

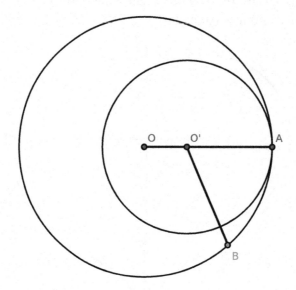

Figure 3.2.16: [3.2, Cor. 2]

For, since two circles cannot have more than two points common [3.10] and their point of intersection is equivalent to two points for purposes of calculation, circles that touch cannot have any other point common. The following is a formal proof of this Corollary:

Construct $\odot O$ and $\odot O'$ where A is the point of intersection, and let O' lie between O and A. Take any other point B in the circumference of $\odot O$, and construct $\overline{O'B}$. By [3.7], $\overline{O'B} > \overline{O'A}$. Therefore, B is outside the circumference of the inner circle. Hence, B cannot be common to both circles. Since point B was chosen arbitrarily, the circles cannot have any other common point except for A.

3.2. PROPOSITIONS FROM BOOK III

Proposition 3.13. *INTERSECTING CIRCLES II.*

Two circles cannot touch each other at two points either internally or externally.

Proof. We divide the proof into its internal and external cases:

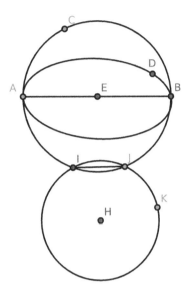

Figure 3.2.17: [3.13]

Internal case: suppose two distinct circles $\odot ACB$ and $\odot ADB$ touch internally at two points A and B. Since the two circles touch at A, the segment joining their centers passes through A [3.11]. Similarly, the segment joining their centers passes through B. Hence, the centers of these circles and the points A and B are on $\overline{AB}$, and so $\overline{AB}$ is a diameter of each circle. Bisect $\overline{AB}$ at E: clearly, E is the center of each circle, i.e., the circles are concentric. This contradicts [3.5], and thus $\odot ACB$ and $\odot ADB$ do not touch internally at two points.

External case: if two circles $\odot E$ and $\odot H$ touch externally at points I and J where I and J are distinct, then by [3.12] $\overline{EH}$ contains the points I and J; in other words, I and J are not distinct, a contradiction. Thus, $\odot E$ and $\odot H$ do not touch externally at two points. □

An alternative proof to the internal case:

Proof. Construct a line bisecting $\overline{AB}$ perpendicularly. By [3.1, Cor. 1], this line passes through the center of each circle, and by [3.11] and [3.12] must pass through each point of intersection, a contradiction. Hence, two circles cannot touch each other at two points. □

Remark. This proposition is an immediate inference from [3.12, Cor. 1] that if a point of intersection counts for two intersections, then two contacts would be equivalent to four intersections; but there cannot be more than two intersections [3.10]. It also follows from [3.12, Cor. 2] that if two circles touch each other at point A, they cannot have any other point in common; hence, they cannot touch again at B.

Exercises.

1. If a circle with a non-fixed center touches two fixed circles externally, the difference between the distances of its center from the centers of the fixed circles is equal to the difference or the sum of their radii, according to whether the contacts are of the same or of opposite type [Def. 3.4].

2. If a circle with a non-fixed center is touched by one of two fixed circles internally and touches the other fixed circle either externally or internally, the sum of the distances from its center to the centers of the fixed circles is equal to the sum or the difference of their radii, according to whether the contact with the second circle is internal or external.

3. Suppose two circles touch externally. If through the point of intersection any secant is constructed cutting the circles again at two points, the radii constructed to these points are parallel. [See the final chapter for a solution.]

4. Suppose two circles touch externally. If two diameters in these circles are parallel, the line from the point of intersection to the endpoint of one diameter passes through the endpoint of the other. [See the final chapter for a solution.]

3.2. PROPOSITIONS FROM BOOK III

Proposition 3.14. *EQUALITY OF CHORD LENGTHS.*

Chords in a circle are equal in length if and only if they are equally distant from the center.

Proof. Construct $\odot O$ with chords $\overline{AB}$ and $\overline{CD}$. We claim that $\overline{AB} = \overline{CD}$ if and only if $\overline{AB}$ and $\overline{CD}$ are equally distant from the center.

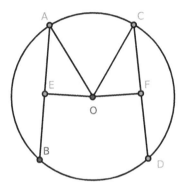

Figure 3.2.18: [3.14]

Suppose $\overline{AB} = \overline{CD}$, and construct $\overline{OE} \perp \overline{AB}$ and $\overline{OF} \perp \overline{CD}$. We claim that $\overline{EO} = \overline{FO}$.

Construct $\overline{AO}$ and $\overline{CO}$. Because $\overline{AB}$ is a chord in $\odot O$ and $\overline{OE}$ is a perpendicular segment constructed from the center to E, $\overline{OE}$ bisects $\overline{AB}$ [3.3]; or $\overline{AE} = \overline{EB}$. Similarly, $\overline{CF} = \overline{FD}$. Since $\overline{AB} = \overline{CD}$ by hypothesis, $\overline{AE} = \overline{CF}$.

Because $\angle OEF$ is a right angle, $(\overline{AO})^2 = (\overline{AE})^2 + (\overline{EO})^2$ by [1.47]. Similarly, $(\overline{CO})^2 = (\overline{CF})^2 + (\overline{FO})^2$. Since $(\overline{AO})^2 = (\overline{CO})^2$ and $(\overline{AE})^2 = (\overline{CF})^2$, we have $(\overline{EO})^2 = (\overline{FO})^2$, and so $\overline{EO} = \overline{FO}$.

Now suppose $\overline{EO} = \overline{FO}$ under a construction similar to the above. We claim that $\overline{AB} = \overline{CD}$.

Construct $\overline{AO}$ and $\overline{CO}$. By [1.47] and similarly to the above proof, $(\overline{AE})^2 + (\overline{EO})^2 = (\overline{CF})^2 + (\overline{FO})^2$ where $(\overline{EO})^2 = (\overline{FO})^2$ due to our hypothesis. Hence $(\overline{AE})^2 = (\overline{CF})^2$, and so $\overline{AE} = \overline{CF}$. But $\overline{AB} = 2 \cdot \overline{AE}$ and $\overline{CD} = 2 \cdot \overline{CF}$ by [3.3], and so $\overline{AB} = \overline{CD}$. □

Exercises.

1. If a chord of given length slides around a fixed circle, then:

    (a) the locus of its midpoint is a circle;

    (b) the locus of any point fixed on the chord is a circle.

Proposition 3.15. *INEQUALITY OF CHORD LENGTHS.*

The diameter is the longest chord in a circle, and a chord is nearer to the center of a circle than another chord if and only if it is the longer of the two chords.

Proof. Construct $\odot O$ with diameter $\overline{AB}$ and chords $\overline{CD}$ and $\overline{EF}$ such that $\overline{CD}$ is nearer to O that $\overline{EF}$. We claim that:

(1) $\overline{AB}$ is the longest chord in a circle;

(2) $\overline{CD} > \overline{EF}$;

(3) Longer chords are nearer to the center than shorter chords.

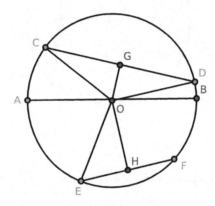

Figure 3.2.19: [3.15]

(1) We claim that $\overline{AB}$ is the longest chord in a circle.

Construct $\overline{OC}, \overline{OD}, \overline{OE}$ as well as $\overline{OG} \perp \overline{CD}$ and $\overline{OH} \perp \overline{EF}$. Notice that $\overline{AB} = \overline{OA} + \overline{OB} = \overline{OC} + \overline{OD}$. Consider $\triangle CDO$: $\overline{OC} + \overline{OD} > \overline{CD}$ by [1.20]. Therefore, $\overline{AB} > \overline{CD}$. Since the choice of $\overline{CD}$ was arbitrary, the proof is complete.

(2) We claim that $\overline{CD} > \overline{EF}$.

Since $\overline{CD}$ is nearer to O than $\overline{EF}$ by hypothesis, it follows that $\overline{OG} < \overline{OH}$ [3.14]. Since $\triangle OGC$ and $\triangle OHE$ are right triangles, we find that $(\overline{OC})^2 = (\overline{OG})^2 + (\overline{GC})^2$ and $(\overline{OE})^2 = (\overline{OH})^2 + (\overline{HE})^2$. Since $\overline{OC} = \overline{OE}$, $(\overline{OG})^2 + (\overline{GC})^2 = (\overline{OH})^2 + (\overline{HE})^2$. But $(\overline{OG})^2 < (\overline{OH})^2$, and so $(\overline{GC})^2 > (\overline{HE})^2$. It follows that $\overline{GC} > \overline{HE}$. Since $\overline{CD} = 2 \cdot \overline{GC}$ and $\overline{EF} = 2 \cdot \overline{HE}$ by [3.3], $\overline{CD} > \overline{EF}$.

(3) Longer chords are nearer to the center than shorter chords.

Suppose that $\overline{CD} > \overline{EF}$. We claim that $\overline{OG} < \overline{OH}$.

As before, we find that $(\overline{OG})^2 + (\overline{GC})^2 = (\overline{OH})^2 + (\overline{HE})^2$. Due to our hypothesis, $(\overline{GC})^2 > (\overline{HE})^2$. Therefore $(\overline{OG})^2 < (\overline{OH})^2$, and so $\overline{OG} < \overline{OH}$. □

Exercises.

1. The shortest chord which can be constructed through a given point within a circle is the perpendicular to the diameter which passes through that point.

2. Through a given point, within or outside of a given circle, construct a chord of length equal to that of a given chord.

3. Through one of the points of intersection of two circles, construct a secant

    (a) where the sum of its segments intercepted by the circles is a maximum;

    (b) which is of any length less than that of the maximum.

4. Suppose that circles touch each other externally at A, B, C and that the chords $\overline{AB}$, $\overline{AC}$ of two of them are extended to meet the third again in the points D and E. Prove that $\overline{DE}$ is a diameter of the third circle and is parallel to the segment joining the centers of the others.

Proposition 3.16. *THE PERPENDICULAR TO A DIAMETER OF A CIRCLE.*

The perpendicular to a diameter of a circle intersects the circumference at one and only one point, and any other segment through the diameter's endpoint intersects the circle at two points.

Proof. Construct $\odot C$ with points A, B, and H on its circumference where $\overline{AB}$ is a diameter of $\odot C$. Also construct $\overline{BH}$ and $\overleftrightarrow{BI} \perp \overline{AB}$ where $\overleftrightarrow{BI}$ intersects $\odot C$ at B. We claim that:

(1) $\overleftrightarrow{BI}$ touches $\odot C$ at B only;

(2) $\overline{BH}$ cuts $\odot C$.

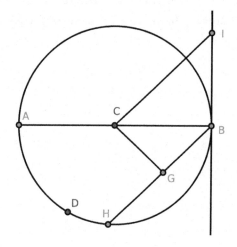

Figure 3.2.20: [3.16]

Claim 1: $\overleftrightarrow{BI}$ touches $\odot C$ at B only.

Let I be an arbitrary point on $\overleftrightarrow{BI}$ and construct the segment $\overline{CI}$. Because $\angle CBI$ is a right angle, $(\overline{CI})^2 = (\overline{CB})^2 + (\overline{BI})^2$ by [1.47]. It follows that $(\overline{CI})^2 > (\overline{CB})^2$, and so $\overline{CI} > \overline{CB}$. By [3.2], I lies outside of $\odot C$. Similarly, every other point on $\overleftrightarrow{BI}$ except B lies outside of the $\odot C$. Hence, $\overleftrightarrow{BI}$ intersects touches $\odot C$ at B only.

Claim 2: $\overline{BH}$ cuts $\odot C$.

Construct $\overline{CG} \perp \overline{BH}$. It follows that $(\overline{BC})^2 = (\overline{CG})^2 + (\overline{GB})^2$. Therefore $(\overline{BC})^2 > (\overline{CG})^2$, and so $\overline{BC} > \overline{CG}$. By [3.2], G must lie within $\odot C$, and consequently if $\overline{BG}$ is extended it must also intersect $\odot C$ at H and therefore cut it.

This completes the proof. □

Exercises.

1. If two circles are concentric, all chords of the larger circle which touch the smaller circle are equal in length. [See the final chapter for a solution.]

2. Construct a parallel to a given line which touches a given circle.

3. Construct a perpendicular to a given line which touches a given circle.

4. Construct a circle having its center at a given point

   (a) and touches a given line;

   (b) and touches a given circle.

How many solutions exist in this case?

5. Construct a circle of given radius that touches two given lines. How many solutions exist?

6. Find the locus of the centers of a system of circles touching two given lines.

7. Construct a circle of given radius that touches a given circle and a given line or that touches two given circles.

Proposition 3.17. *TANGENTS ON CIRCLES I.*

It is possible to construct a tangent of a given circle from a given point outside of the circle.

Proof. Construct $\odot O$ and P such that P is outside of $\odot O$. We wish to construct tangent $\overline{BP}$ to $\odot O$.

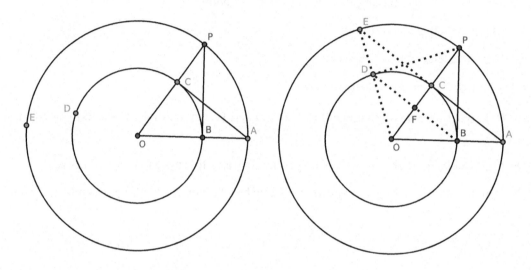

Figure 3.2.21: [3.17] (α), (β)

Construct radius $\overline{OC}$ and extend $\overline{OC}$ to $\overline{OP}$. With O as center and $\overline{OP}$ as radius, construct the circle $\odot O_1$. Also construct $\overline{CA} \perp \overline{OP}$. Construct $\overline{OA}$, intersecting $\odot O$ at B, and construct $\overline{BP}$. We claim that $\overline{BP}$ is the required tangent to $\odot O$.

Since O is the center of $\odot O$ and $\odot O_1$, we find that $\overline{OA} = \overline{OP}$ and $\overline{OC} = \overline{OB}$. Consider $\triangle AOC$ and $\triangle POB$: $\overline{OA} = \overline{OP}$, $\overline{OC} = \overline{OB}$, and each shares $\angle BOC = \angle AOC = \angle POB$. By [1.4], $\triangle AOC \cong \triangle POB$, and so $\angle OCA = \angle OBP$.

But $\angle OCA$ is a right angle by construction; therefore $\angle OBP$ is a right angle, and by [3.16], $\overline{BP}$ touches the circle $\odot O$ at B. By definition $\overline{BP}$ is a tangent of $\odot O$ at point B, which proves our claim. □

Corollary. 3.17.1. If $\overline{AC}$ is extended to $\overline{AE}$ and $\overline{OE}$ is constructed, then $\odot O$ is cut at D. Construct $\overline{DP}$; $\overline{DP}$ is a second tangent of $\odot O$ at P.

3.2. PROPOSITIONS FROM BOOK III

Exercises.

1. Prove that tangents $\overline{PB}$ and $\overline{PD}$ in [3.17] are equal in length because the square of each is equal to the square of $\overline{OP}$ minus the square of the radius.

2. If a quadrilateral is circumscribed to a circle, the sum of the lengths of one pair of opposite sides is equal to the sum of the lengths of the other pair.

3. If a parallelogram is circumscribed to a circle, it must be a rhombus, and so its diagonals intersect at the center.

4. If $\overline{BD}$ is constructed and $\overline{OP}$ is intersected at F, then $\overline{OP} \perp \overline{BD}$.

5. The locus of the intersection of two equal tangents to two circles is a segment (called the *radical axis* of the two circles).

6. Find a point such that tangents from it to three given circles is equal. (This point is called the *radical center* of the three circles.)

7. Prove that the rectangle $OF \cdot OP$ is equal in area to the square of the radius of $\odot O$. (Note: we are locating the inverse points with respect to $\odot O$. See the definition below.)

8. The intercept made on a variable tangent by two fixed tangents stands opposite a constant angle at the center.

9. Construct a common tangent to two circles. Demonstrate how to construct a segment cutting two circles so that the intercepted chords are of given lengths.

10. Prove [Cor. 3.17.1].

Proposition 3.18. *TANGENTS ON CIRCLES II.*

If a line touches a circle, the segment from the center of the circle to the point of intersection with the line is perpendicular to the line.

Proof. Construct $\odot O$ with point C on its circumference, and also construct $\overleftrightarrow{CD}$. We claim that if $\overleftrightarrow{CD}$ touches $\odot O$, then $\overline{OC} \perp \overleftrightarrow{CD}$.

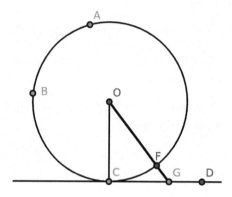

Figure 3.2.22: [3.18]

Suppose instead that another segment $\overline{OG}$ is constructed from the center such that $\overline{OG} \perp \overleftrightarrow{CD}$ where $\overline{OG}$ cuts the circle at F. Because the angle $\angle OGC$ is right by hypothesis, the angle $\angle OCG$ must be acute [1.17]. By [1.19], $\overline{OC} > \overline{OG}$. But $\overline{OG} = \overline{OF} \oplus \overline{FG}$ and $\overline{OC} = \overline{OF}$, and so $\overline{OC} < \overline{OG}$, a contradiction. Hence $\overline{OC} \perp \overleftrightarrow{CD}$. □

Alternatively:

Proof. Since the perpendicular must be the shortest segment from O to $\overleftrightarrow{CD}$ and $\overline{OC}$ is evidently the shortest line, it follows that $\overline{OC} \perp \overleftrightarrow{CD}$. □

3.2. PROPOSITIONS FROM BOOK III

Proposition 3.19. *TANGENTS ON CIRCLES III.*

If a line is a tangent to a circle, then the perpendicular constructed from its point of intersection passes through the center of the circle.

Proof. Suppose $\overleftrightarrow{AB}$ is tangent to $\odot CDA$. We claim that if $\overleftrightarrow{AB} \perp \overline{AC}$, then $\overline{AC}$ contains the center of $\odot CDA$.

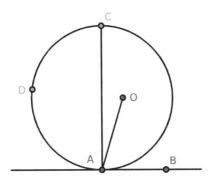

Figure 3.2.23: [3.19]

Suppose otherwise: let O be the center of $\odot CDA$ and construct $\overline{AO}$. Notice that $\angle OAC > 0$. Because $\overleftrightarrow{AB}$ touches $\odot CDA$ and $\overline{OA}$ is constructed from the center to the point of intersection, $\overline{OA} \perp \overleftrightarrow{AB}$ by [3.18]. Since $\overleftrightarrow{AB} \perp \overline{AC}$ by hypothesis, $\angle OAB$ and $\angle CAB$ are right angles.

It follows that $\angle CAB = \angle OAB$ and $\angle CAB = \angle OAB + \angle OAC$; hence $\angle OAC = 0$ and $\angle OAC > 0$, a contradiction. Therefore, the center lies on $\overline{AC}$. □

Corollary. 3.19.1. *If a number of circles touch the same line at the same point, the locus of their centers is the perpendicular to the line at the point.*

Corollary. 3.19.2. *Suppose we have a circle and any two of the following properties:*

(a) a tangent to a circumference;

(b) a segment, ray, or straight line constructed from the center of the circle to the point of intersection;

(c) right angles at the point of intersection.

Then by [3.16], [3.18], [3.19], and the Rule of Symmetry, the remaining property follows. If we have (a) and (c), then it may be necessary to extend a given segment or a ray to the center of the circle: these are limiting cases of [3.1, Cor. 1] and [3.3].

Proposition 3.20. *ANGLES AT THE CENTER OF A CIRCLE AND ON THE CIRCUMFERENCE.*

The angle at the center of a circle is double the angle at the circumference when each stands on the same arc of the circumference.

Proof. Construct $\odot E$ with $\overline{EB}$ as radius, and construct $\angle BEC$ and $\angle BAC$ where A, C, and D are points on the circumference of $\odot E$. We claim that $\angle BEC = 2 \cdot \angle BAC = 2 \cdot \angle BDC$.

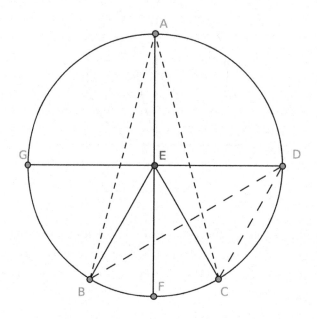

Figure 3.2.24: [3.20]

Construct $\overline{AEF}$, and consider $\triangle EAB$: since $\overline{EA} = \overline{EB}$, by [1.6] $\angle EAB = \angle EBA$ and therefore $\angle EAB + \angle EBA = 2 \cdot \angle EAB$. Since $\angle BEF = \angle EAB + \angle EBA$ by [1.32], $\angle BEF = 2 \cdot \angle EAB$. Similarly in $\triangle EAC$, $\angle FEC = 2 \cdot \angle EAC$. It follows that

$$\angle BEC = \angle BEF + \angle FEC = 2 \cdot (\angle EAB + \angle EAC) = 2 \cdot \angle BAC$$

Now construct $\overline{GD}$, $\overline{BD}$, and $\overline{CD}$. By an argument similar to the above, we can prove that $\angle GEC = 2 \cdot \angle EDC$ and $\angle GEB = 2 \cdot \angle EDB$. Since $\angle BEC = \angle GEC - \angle GEB$, we find that

$$\angle BEC = 2 \cdot (\angle EDC - \angle EDB) = 2 \cdot \angle BDC$$

which completes the proof. $\square$

Corollary. *3.20.1. The angle in a semicircle is a right angle.*

3.2. PROPOSITIONS FROM BOOK III

Proposition 3.21. *ANGLES ON CHORDS.*

In a circle, angles standing on the same arc are equal in measure to each other.

Proof. Construct $\circ E$, and also construct $\angle BAC$ and $\angle BDC$ on the same arc BFC. We claim that $\angle BAC = \angle BDC$.

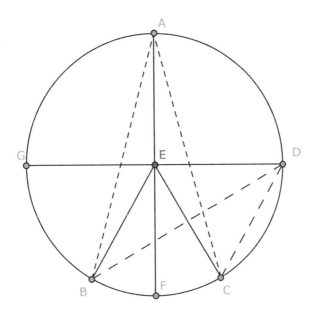

Figure 3.2.25: [3.21]

By [3.20], $\angle BEC = 2 \cdot \angle BAC = 2 \cdot \angle BDC$, or $\angle BAC = \angle BDC$. This proves our claim. □

Corollary. 3.21.1. *If two triangles $\triangle ACB$, $\triangle ADB$ stand on the same base $\overline{AB}$ and have equal vertical angles on the same side of it, then the four points A, C, D, B are concyclic.*

Corollary. 3.21.2. *If A, B are two fixed points and if C varies its position in such a way that the angle $\angle ACB$ retains the same value throughout, the locus of C is a circle. (Or: given the base of a triangle and the vertical angle, the locus of the vertex is a circle).*

Exercises.

1. Given the base of a triangle and the vertical angle, find the locus

    (a) of the intersection of its perpendiculars;

(b) of the intersection of the internal bisectors of its base angles;

(c) of the intersection of the external bisectors of the base angles;

(d) of the intersection of the external bisector of one base angle and the internal bisector of the other.

2. If the sum of the squares of two segments is given, prove that their sum is a maximum when the segments are equal in length.

3. Of all triangles having the same base and vertical angle, prove that the sum of the sides of an isosceles triangle is a maximum.

4. Of all triangles inscribed in a circle, the equilateral triangle has the maximum perimeter.

5. Of all concyclic figures having a given number of sides, the area is a maximum when the sides are equal.

6. Prove [Cor. 3.20.1].

7. Prove [Cor. 3.21.1].

8. Prove [Cor. 3.21.2].

3.2. PROPOSITIONS FROM BOOK III

Proposition 3.22. *QUADRILATERALS INSCRIBED INSIDE CIRCLES.*

The sum of the opposite angles of a quadrilateral inscribed in a circle equals two right angles.

Proof. Construct quadrilateral $ABCD$ inscribed in $\odot E$. We claim that the sum of the opposite angles of $ABCD$ equals two right angles.

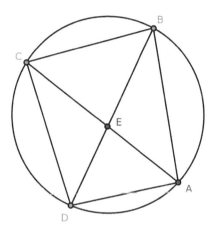

Figure 3.2.26: [3.22]

Let π radians = two right angles, and construct diameters $\overline{AC}$ and $\overline{BD}$. Since $\angle ABD$ and $\angle ACD$ stand on arc AD, $\angle ABD = \angle ACD$ by [3.21]. Similarly, $\angle DBC = \angle DAC$ because they stand on arc DC. Hence

$$\begin{aligned} \angle ABC &= \angle ABD + \angle DBC \\ &= \angle ACD + \angle DAC \end{aligned}$$

From this, we obtain

$$\angle ABC + \angle CDA = \angle ACD + \angle DAC + \angle CDA$$

where the right-hand side of the equality is the sum of the interior angles of $\triangle ACD$. Since this sum equals π radians by [1.32], $\angle ABC + \angle CDA = \pi$ radians.

Similarly, $\angle DAB + \angle BCD = \pi$ radians (*mutatis mutandis*), which proves our claim. □

Corollary. 3.22.1. *If the sum of two opposite angles of a quadrilateral are equal to two right angles, then a circle may be inscribed about the quadrilateral.*

Corollary. 3.22.2. *If a parallelogram is inscribed in a circle, then it is a rectangle.*

Exercises.

1. If the opposite angles of a quadrilateral are supplemental, it is cyclic.

2. A segment which makes equal angles with one pair of opposite sides of a cyclic quadrilateral makes equal angles with the remaining pair and with the diagonals.

3. If two opposite sides of a cyclic quadrilateral are extended to meet and a perpendicular falls on the bisector of the angle between them from the point of intersection of the diagonals, this perpendicular will bisect the angle between the diagonals.

4. If two pairs of opposite sides of a cyclic hexagon are respectively parallel to each other, the remaining pair of sides are also parallel.

5. If two circles intersect at the points A, B, and any two segments $\overline{ACD}$, $\overline{BFE}$ are constructed through A and B, cutting one of the circles in the points C, E and the other in the points D, F, then $CE \parallel DF$.

6. If equilateral triangles are constructed on the sides of any triangle, the segments joining the vertices of the original triangle to the opposite vertices of the equilateral triangles are concurrent.

7. In the same case as #7, prove that the centers of the circles constructed about the equilateral triangles form another equilateral triangle.

8. If a quadrilateral is constructed about a circle, the angles at the center standing opposite the opposite sides are supplemental.

9. If a tangent which varies in position meets two parallel tangents, it stands opposite a right angle at the center.

10. If a hexagon is circumscribed about a circle, the sum of the angles standing opposite the center from any three alternate sides is equal to two right angles.

11. Prove [Cor. 3.22.1].

12. After completing #11, rewrite the results of [3.22] and [Cor. 3.22.1] into one proposition.

13. Prove [Cor. 3.22.2].

3.2. PROPOSITIONS FROM BOOK III

Proposition 3.23. *UNIQUENESS OF ARCS.*

Two similar and unequal arcs cannot be constructed on the same side of the same chord.

Proof. Construct $\overline{AB}$. Suppose instead that two similar and unequal arcs ACB and ADB are constructed on the same side of $\overline{AB}$. Construct $\overline{ADC}, \overline{CB},$ and $\overline{DB}$.

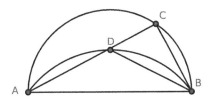

Figure 3.2.27: [3.23]

Since arc ACB is similar to arc ADB, $\angle ADB = \angle ACB$ by [Def. 3.10]; this contradicts [1.16] and proves our claim. □

Proposition 3.24. *EQUALITY OF SIMILAR ARCS.*

Similar arcs standing on equal chords are equal in length.

Proof. Construct $\overline{AB} = \overline{CD}$ and arcs AEB and CFD such that $AEB \sim CFD$. We claim $AEB = CFD$.

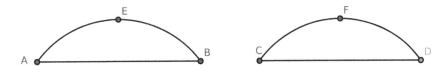

Figure 3.2.28: [3.24]

Since $\overline{AB} = \overline{CD}$, if $\overline{AB}$ is applied to $\overline{CD}$ such that the point A coincides with C and the point B coincides with D, then the chord $\overline{AB}$ coincides with $\overline{CD}$. Because $AEB \sim CFD$, they must coincide at every point [3.23]. This proves our claim. □

Corollary. *3.24.1. Since the chords are equal in length, they are congruent; therefore the arcs, being similar, are also congruent.*

Proposition 3.25. *CONSTRUCTION OF A CIRCLE FROM AN ARC.*

Given an arc of a circle, it is possible to construct the circle to which the arc belongs.

Proof. Given an arc (ABC) of $\bigcirc F$, we wish to construct $\bigcirc F$.

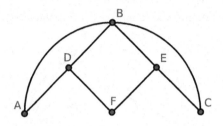

Figure 3.2.29: [3.25]

Take any three points A, B, C on arc ABC. Construct $\overline{AB}$, $\overline{BC}$. Bisect $\overline{AB}$ at D and $\overline{BC}$ at E. Construct $\overline{DF} \perp \overline{AB}$ and $\overline{EF} \perp \overline{BC}$. We claim that F, the intersection of $\overline{DF}$ and $\overline{EF}$, is the center of the required circle.

Because $\overline{DF}$ bisects and is perpendicular to $\overline{AB}$, the center of the circle of which ABC is an arc lies on $\overline{DF}$ [3.1, Cor. 1]. Similarly, the center of the circle of which ABC is an arc lies on $\overline{EF}$.

Since F is the intersection of $\overline{DF}$ and $\overline{EF}$, our proof is complete. □

3.2. PROPOSITIONS FROM BOOK III

Proposition 3.26. *ANGLES AND ARCS I.*

In equal circles, equal angles at the centers or on the circumferences stand upon arcs of equal length.

Proof. Construct $\odot G$ and $\odot H$ with equal radii and with equal angles at the centers: $\angle BGC$ in $\odot G$ and $\angle EHF$ in $\odot H$. Also construct equal angles $\angle BAC$ in $\odot G$ where A is on the circumference and $\angle EDF$ in $\odot H$ where D is on the circumference. We claim that arc $BKC =$ arc ELF.

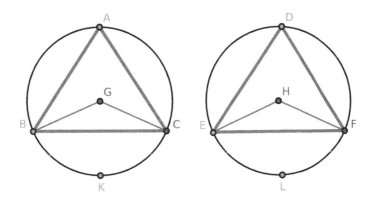

Figure 3.2.30: [3.26]

Construct $\overline{BC}$ and $\overline{EF}$ and consider $\triangle BGC$ and $\triangle EHF$: $\overline{BG} = \overline{EH}$, $\overline{GC} = \overline{HF}$, and $\angle BGC = \angle EHF$. By [1.4], $\triangle BGC \cong \triangle EHF$, and so $\overline{BC} = \overline{EF}$.

By [3.20], $\angle BAC = \frac{1}{2} \cdot \angle BGC$ and $\angle EDF = \frac{1}{2} \cdot \angle EHF$. Since $\angle BGC = \angle EHF$ by hypothesis, $\angle BAC = \angle EDF$. By [Def. 3.10], arc $BAC \sim$ arc EDF. And by [3.24], arc $BAC =$ arc EDF.

By [Def. 3.1], $\odot G$ and $\odot H$ are equal in measure, and so arc $BKC =$ arc ELF which proves our claim. □

Corollary. *3.26.1. If the opposite angles of a cyclic quadrilateral are equal, one of its diagonals must be a diameter of the circumscribed circle.*

Corollary. *3.26.2. Parallel chords in a circle intercept equal arcs.*

Corollary. *3.26.3. If two chords intersect at any point within a circle, the sum of the opposite arcs which they intercept is equal to the arc which parallel chords intersecting on the circumference intercept. If two chords intersect at any point outside a circle, the difference of the arcs they intercept is equal to the arc which parallel chords intersecting on the circumference intercept.*

Corollary. *3.26.4. If two chords intersect at right angles, the sum of the opposite arcs which they intercept on the circle is a semicircle.*

Exercises.

1. Prove [Cor. 3.26.1].

2. Prove [Cor. 3.26.2].

3. Prove [Cor. 3.26.3].

4. Prove [Cor. 3.26.4].

3.2. PROPOSITIONS FROM BOOK III

Proposition 3.27. *ANGLES AND ARCS II.*

In equal circles, angles at the centers or at the circumferences which stand on equal arcs are equal in measure.

Proof. Construct $\odot G$ and $\odot H$ with equal radii and construct angles at the centers ($\angle AOB$, $\angle DHE$) and at the circumferences ($\angle ACB$, $\angle DFE$) which stand on equal arcs (AGB, DKE). We claim that $\angle AOB = \angle DHE$ and $\angle ACB = \angle DFE$.

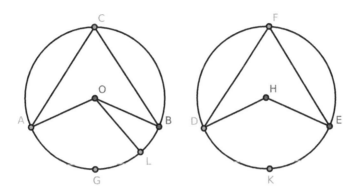

Figure 3.2.31: [3.27]

Consider the angles at the centers ($\angle AOB$, $\angle DHE$). Suppose that $\angle AOB > \angle DHE$ and that $\angle AOL = \angle DHE$. Since the circles are equal in all respects, arc AGL = arc DKE [3.26]. Notice that arc $LB > 0$.

But arc AGB = arc DKE by hypothesis. Hence arc AGB = arc AGL and arc AGB = arc $AGL \oplus$ arc LB where arc $LB = 0$, a contradiction. A corresponding contradiction follows if we assume that $\angle AOB < \angle DHE$. Therefore, $\angle AOB = \angle DHE$.

Now consider the angles at the circumference. Since $2 \cdot \angle ACB = \angle AOB$ and $2 \cdot \angle DFE = \angle DHE$ by [3.20], $\angle ACB = \angle DFE$. This completes the proof. □

Proposition 3.28. *CHORDS AND ARCS I.*

In equal circles, chords of equal length divide the circumferences into arcs, and these arcs are respectively equal.

Proof. Construct equal circles ($\odot O$, $\odot H$) with equal chords ($\overline{AB}$, $\overline{DE}$). We claim that $\overline{AB}$ and $\overline{DE}$ divide the circumferences of $\odot O$ and $\odot H$, respectively, so that arc AGB = arc DKE and arc ACB = arc DFE.

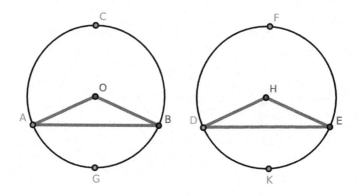

Figure 3.2.32: [3.28]

If the equal chords are diameters, all arcs are equal semicircles, which completes the proof.

Otherwise, construct $\overline{AO}$, $\overline{OB}$, $\overline{DH}$, and $\overline{HE}$. Because the circles are equal in all respects, their radii are equal [Def. 3.1].

Consider $\triangle AOB$ and $\triangle DHE$: $\overline{AO} = \overline{DH}$, $\overline{OB} = \overline{HE}$, and $\overline{AB} = \overline{DE}$. By [1.8], $\triangle AOB \cong \triangle DHE$; hence $\angle AOB = \angle DHE$, and so arc AGB = arc DKE [3.26].

Since the whole circumference $AGBC$ is equal in measure to the whole circumference $DKEF$, it follows that arc ACB = arc DFE. This proves our claim. □

3.2. PROPOSITIONS FROM BOOK III

Proposition 3.29. *CHORDS AND ARCS II.*

In equal circles, equal arcs stand opposite equal chords.

Proof. Construct equal circles $\odot O$ and $\odot H$ where arc $AGB =$ arc DKE. We claim that $\overline{AB} = \overline{DE}$.

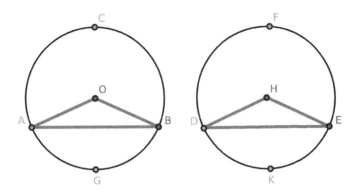

Figure 3.2.33: [3.29]

Construct $\overline{AO}, \overline{OB}, \overline{DH}$, and $\overline{HE}$. Because the circles are equal in measure, the angles $\angle AOB$ and $\angle DHE$ at the centers which stand on the equal arcs AGB and DKE are themselves equal [3.27].

Consider $\triangle AOB$ and $\triangle DHE$: $\overline{AO} = \overline{DH}, \overline{OB} = \overline{HE}$, and $\angle AOB = \angle DHE$. By [1.4], $\triangle AOB \cong \triangle DHE$. Therefore, $\overline{AB} = \overline{DE}$, which proves our claim. □

Corollary. *3.29.1. Propositions [3.26]-[3.29] are related in the following sense: in circles with equal radius,*

1. In [3.26], equal angles imply equal arcs.

2. In [3.27], equal arcs imply equal angles. Together, [3.26] and [3.27] state that equal angles and equal arcs are equivalent.

3. In [3.28], equal chords imply equal arcs.

4. In [3.29], equal arcs imply equal chords. Together, [3.28] and [3.29] state that equal chords and equal arcs are equivalent.

Or: in circles with equal radius, equal chords $\iff$ equal angles $\iff$ angles stand on equal arcs.

Remark. Since the two circles in the four last propositions are equal, they are congruent figures, and the truth of the propositions is made evident by superposition.

Proposition 3.30. *BISECTING AN ARC.*

Proof. We wish to bisect the given arc ACB.

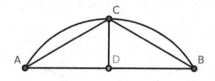

Figure 3.2.34: [3.30]

Construct the chord $\overline{AB}$ and bisect it at D. Construct $\overline{DC} \perp \overline{AB}$, intersecting the arc at C. We claim that arc ACB is bisected at C.

Construct $\overline{AC}$ and $\overline{BC}$, and consider $\triangle ADC$ and $\triangle BDC$: $\overline{AD} = \overline{DB}$, they share side $\overline{DC}$, and $\angle ADC = \angle BDC$. By [1.4], $\triangle ADC \cong \triangle BDC$, and so $\overline{AC} = \overline{BC}$.

By [3.28], arc AC = arc BC; since arc ACB = arc $AC \oplus$ arc BC, arc ACB is bisected at C. □

Exercises.

1. Suppose that $ABCD$ is a semicircle with diameter $\overline{AD}$ and a chord $\overline{BC}$. Extend $\overline{BC}$ to $\overrightarrow{BC}$ and $\overline{AD}$ to $\overrightarrow{AD}$, and suppose each ray intersects at E. Prove that if $\overline{CE}$ is equal in length to the radius of $ABCD$, then arc $\overline{AB} = 3 \cdot \overline{CD}$. [See the final chapter for a solution.]

2. The internal and the external bisectors of the vertical angle of a triangle inscribed in a circle meet the circumference again at points equidistant from the endpoints of the base.

3. If A is one of the points of intersection of two given circles and two chords $\overline{ACD}$, $\overline{AC'D'}$ are constructed, cutting the circles in the points C, D, C', and D', then the triangles $\triangle BCD$, $\triangle BC'D'$ formed by joining these to the second point B of intersection of the circles are equiangular.

4. If the vertical angle $\angle ACB$ of a triangle inscribed in a circle is bisected by a line $\overleftrightarrow{CD}$ which meets the circle again at D, and from D perpendiculars $\overline{DE}$, $\overline{DF}$ are constructed to the sides, one of which is extended, prove that $\overline{EA} = \overline{BF}$ and hence that $\overline{CE} = \frac{1}{2}(\overline{AC} + \overline{BC})$.

3.2. PROPOSITIONS FROM BOOK III

Proposition 3.31. *THALES' THEOREM.*

In a circle,

(1) if a circle is divided into two semicircles, then the angle contained in either arc is a right angle;

(2) if a circle is divided into two unequal arcs and an angle is contained in the larger of the two arcs, then the angle contained in that arc is an acute angle;

(3) if a circle is divided into two unequal arcs and an angle is contained in the smaller of the two arcs, then the angle contained in that arc is an obtuse angle.

Proof. Construct $\odot O$, diameter $\overline{AB}$, and points C, D, and E such that C and D are on one semicircle and E is on the other semicircle. We claim that:

(1) $\angle ACB$ in arc ACB is a right angle;

(2) $\angle ACE$ in arc ACE is an acute angle.

(3) $\angle ACD$ in arc ACD is an obtuse angle.

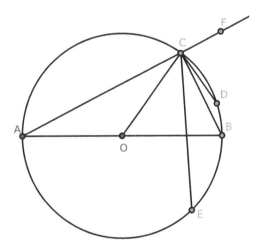

Figure 3.2.35: [3.31]

Claim 1: $\angle ACB$ in arc ACB is a right angle;

Construct radius $\overline{OC}$ and extend $\overline{AC}$ to $\overrightarrow{AC}$. Consider $\triangle AOC$: since $\overline{AO} = \overline{OC}$, $\angle ACO = \angle OAC$. Similarly in $\triangle OCB$, $\angle OCB = \angle CBO$. Hence,

$$\begin{aligned} \angle ACB &= \angle ACO + \angle OCB \\ &= \angle OAC + \angle CBO \\ &= \angle BAC + \angle CBA \end{aligned}$$

Consider $\triangle ABC$: by [1.32], $\angle FCB = \angle BAC + \angle CBA$. Hence, $\angle ACB = \angle FCB$ where each are adjacent angles; thus, $\angle ACB$ is a right angle.

Claim 2: ∠ACE in arc ACE is an acute angle.

Construct $\overline{CE}$. Since ∠ACB = ∠ACE + ∠BCE, ∠ACB > ∠ACE. But we have proven that ∠ACB is a right angle; thus, ∠ACE is acute.

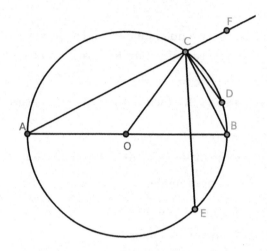

Figure 3.2.36: [3.31]

Claim 3: ∠ACD in arc ACD is an obtuse angle;

Construct $\overline{CD}$. Since ∠ACD = ∠ACB + ∠BCD, ∠ACD > ∠ACB. Since ∠ACB is a right angle, ∠ACE is obtuse. □

Corollary. *3.31.1. If a parallelogram is inscribed in a circle, its diagonals intersect at the center of the circle.*

Corollary. *3.31.2. [3.31] holds if arcs are **replaced** by chords of appropriate length, mutatis mutandis.*

3.2. PROPOSITIONS FROM BOOK III

Proposition 3.32. *TANGENT-CHORD ANGLES RELATED TO ANGLES ON THE CIRCUMFERENCE WHICH STAND ON THE CHORD.*

If a line is tangent to a circle, and from the point of intersection a chord is constructed cutting the circle, the angles made by this chord with the tangent are respectively equal to the angles in the alternate arcs of the circle.

Proof. Construct $\overleftrightarrow{EF}$ such that $\overleftrightarrow{EF}$ is tangent to $\bigcirc ABC$ at A. Construct chord $\overline{AC}$; notice that $\overline{AC}$ cuts $\bigcirc ABC$. We claim that the angles made by this chord with the tangent are respectively equal to the angles in the alternate arcs of the circle. We shall prove this in two cases.

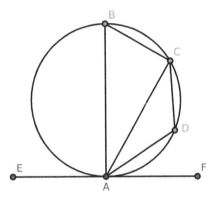

Figure 3.2.37: [3.32](α)

Case 1: We wish to show that $\angle ABC = \angle FAC$ in figure (α).

Construct $\overline{AB}$ such that $\overline{AB} \perp \overleftrightarrow{EF}$. Also construct $\overline{BC}$. Because $\overleftrightarrow{EF}$ is tangent to $\bigcirc ABC$, $\overline{AB}$ is clearly constructed at A, and $\overline{AB} \perp \overleftrightarrow{EF}$, by [3.19] we find that $\overline{AB}$ passes through the center of $\bigcirc ABC$. By [3.31], $\angle ACB$ is a right angle; since $\triangle ABC$ is a triangle, the sum of the two remaining angles, $\angle ABC + \angle CAB$, equals one right angle.

Since $\angle BAF$ is a right angle by construction, $\angle ABC + \angle CAB = \angle BAF$. From this we obtain $\angle ABC = \angle BAF - \angle CAB = \angle FAC$, which proves the first case.

Case 2: We wish to prove that ∠CAE = ∠CDA in figure (β).

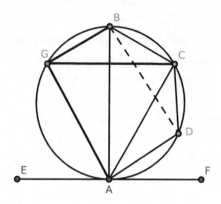

Figure 3.2.38: [3.32](β)

Take any point D on the arc AC. Since the quadrilateral $ABCD$ is inscribed in a circle, the sum of the opposite angles ∠ABC + ∠CDA equals two right angles [3.22] and is therefore equal to the sum ∠FAC+∠CAE. However, ∠ABC = ∠FAC by case 1. Hence, ∠CDA = ∠CAE.

This proves our second and last case, completing the proof. □

Exercises.

1. If two circles touch, any line constructed through the point of intersection will cut off similar segments.

2. If two circles touch and any two lines are constructed through the point of intersection (cutting both circles again), the chord connecting their points of intersection with one circle is parallel to the chord connecting their points of intersection with the other circle.

3. Suppose that ACB is an arc of a circle, $\overleftrightarrow{CE}$ a tangent at C (meeting the chord $\overline{AB}$ extended to E), and $\overline{AD} \perp \overline{AB}$ where D is a point of $\overline{AB}$. Prove that if $\overline{DE}$ be bisected at C then the arc $\overline{AC} = 2 \cdot \overline{CB}$.

4. If two circles touch at a point A and if $\overline{ABC}$ is a chord through A, meeting the circles at points B and C, prove that the tangents at B and C are parallel to each other, and that when one circle is within the other, the tangent at B meets the outer circle at two points equidistant from C.

5. If two circles touch externally, their common tangent at either side stands opposite a right angle at the point of intersection, and its square is equal to the rectangle contained by their diameters.

3.2. PROPOSITIONS FROM BOOK III

Proposition 3.33. *CONSTRUCTING A SEGMENT OF A CIRCLE ON A LINE WHERE THE SEGMENT CONTAINS AN ANGLE EQUAL TO A GIVEN ANGLE.*

Proof. Construct $\overleftrightarrow{AB}$ and $\angle HGF$. On $\overleftrightarrow{AB}$, we wish to construct a segment of a circle which contains an angle equal to $\angle HGF$.

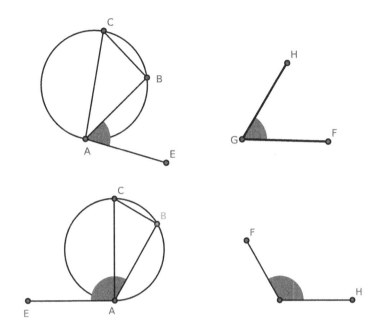

Figure 3.2.39: [3.33]

If $\angle HGF$ is a right angle, construct a semicircle on $\overleftrightarrow{AB}$ as our segment. By [3.31], a semicircle contains a right angle.

Otherwise, construct $\angle BAE = \angle HGF$. Construct $\overline{AC} \perp \overline{AE}$ and $\overline{BC} \perp \overline{AB}$. Let $\overline{AC}$ be the diameter of $\odot ABC$, which we claim is the required circle.

Notice that $\odot ABC$ has B on its circumference because $\angle ABC$ is a right angle [3.31]. Also, $\odot ABC$ touches $\overline{AE}$ because $\angle CAE$ is a right angle [3.16]. It follows that $\angle BAE = \angle ACB$. Since $\angle BAE = \angle HGF$ by construction, $\angle HGF = \angle ACB$.

Thus on $\overleftrightarrow{AB}$, we have constructed a segment of $\odot ABC$ which contains an angle equal to $\angle HGF$. □

Exercises.

1. Construct a triangle, being given the base, vertical angle, and any of the following data:

   (a) a perpendicular.

(b) the sum or difference of the sides.

(c) the sum or difference of the squares of the sides.

(d) the side of the inscribed square on the base.

(e) the median that bisects the base.

2. If lines are constructed from a fixed point to all the points of the circumference of a given circle, prove that the locus of all their points of bisection is a circle.

3. Given the base and vertical angle of a triangle, find the locus of the midpoint of the line joining the vertices of equilateral triangles constructed on the sides.

4. In the same case, find the loci of the vertices of a square constructed on one of the sides.

Proposition 3.34. CONSTRUCTING A SEGMENT OF A CIRCLE WHERE THE SEGMENT CONTAINS AN ANGLE EQUAL TO A GIVEN ANGLE.

Proof. Construct $\odot ABC$ and $\angle HGF$. We wish to construct a segment of $\odot ABC$ which contains an angle equal to $\angle HGF$.

Choose point A on the circumference of $\odot ABC$ and construct the tangent $\overline{AD}$. On $\overline{AD}$, construct $\angle DAC = \angle HGF$. We claim that segment AC of $\odot ABC$ is the required segment.

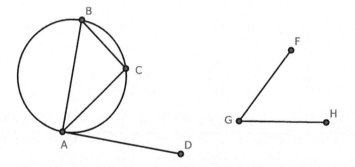

Figure 3.2.40: [3.34]

Choose any point B on the circumference of $\odot ABC$ other than from the arc AC. Construct $\overline{AB}$ and $\overline{BC}$. By [3.32], $\angle DAC = \angle ABC$. But $\angle DAC = \angle HGF$ by construction, and so $\angle HGF = \angle ABC$.

Thus on $\odot ABC$, we have constructed a segment of $\odot ABC$ which contains an angle equal to $\angle HGF$. □

3.2. PROPOSITIONS FROM BOOK III

Proposition 3.35. *AREAS OF RECTANGLES CONSTRUCTED ON CHORDS.*

If two chords of a circle intersect at one point within the circle, then the area of the rectangle contained by the divided segments of the first chord is equal in area to the rectangle contained by the divided segments of the second chord [Def. 2.4].

Proof. Construct $\odot O$ with chords $\overline{AB}$ and $\overline{CD}$ which intersect at one point. We claim that the rectangles contained by the divided segments are equal in area and shall prove this claim in four cases.

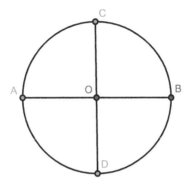

Figure 3.2.41: [3.35], case 1

Case 1: if the point of intersection is the center of $\odot O$, $\overline{AO} = \overline{OB} = \overline{DO} = \overline{OC}$. Hence, $\overline{AO} \cdot \overline{OB} = \overline{DO} \cdot \overline{OC}$.

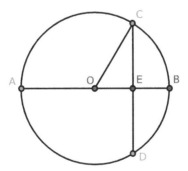

Figure 3.2.42: [3.35], case 2

Case 2: suppose that $\overline{AB}$ passes through O and that $\overline{CD}$ does not; also suppose they intersect at E and that $\overline{AB} \perp \overline{CD}$.

Construct $\overline{OC}$. Because $\overline{AB}$ is divided equally at O and unequally at E, by [2.5]

$$\overline{AE} \cdot \overline{EB} + (\overline{OE})^2 = (\overline{OB})^2$$

Since $\overline{OB} = \overline{OC}$,
$$\overline{AE} \cdot \overline{EB} + (\overline{OE})^2 = (\overline{OC})^2$$
But $(\overline{OC})^2 = (\overline{OE})^2 + (\overline{EC})^2$ by [1.47], and therefore
$$\overline{AE} \cdot \overline{EB} + (\overline{OE})^2 = (\overline{OE})^2 + (\overline{EC})^2$$
$$\overline{AE} \cdot \overline{EB} = (\overline{EC})^2$$

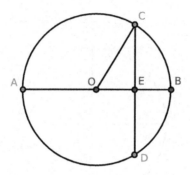

Figure 3.2.43: [3.35], case 2

Because $\overline{AB}$ passes through the center and cuts $\overline{CD}$, which does not pass through the center at a right angle, $\overline{AB}$ bisects $\overline{CD}$ by [3.3]. So $(\overline{EC})^2 = \overline{CE} \cdot \overline{ED}$, and therefore, $\overline{AE} \cdot \overline{EB} = \overline{CE} \cdot \overline{ED}$.

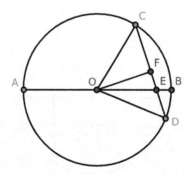

Figure 3.2.44: [3.35], case 3

3. Construct diameter $\overline{AB}$ of $\odot O$ which cuts $\overline{CD}$ such that $\overline{AB} \not\perp \overline{CD}$.

Construct $\overline{OC}$, $\overline{OD}$, and $\overline{OF} \perp \overline{CD}$ [1.11]. Since $\overline{CD}$ is cut at right angles by $\overline{OF}$ where $\overline{OF}$ passes through O, $\overline{CD}$ is bisected at F [3.3] and divided unequally at E. By [2.5], $\overline{CE} \cdot \overline{ED} + (\overline{FE})^2 = (\overline{FD})^2$.

3.2. PROPOSITIONS FROM BOOK III   197

Adding $(\overline{OF})^2$ to each side of the equality and applying [1.47], we obtain:

$$\begin{aligned}\overline{CE}\cdot\overline{ED}+(\overline{FE})^2+(\overline{OF})^2 &= (\overline{FD})^2+(\overline{OF})^2 \\ \overline{CE}\cdot\overline{ED}+(\overline{OE})^2 &= (\overline{OD})^2 \\ \overline{CE}\cdot\overline{ED}+(\overline{OE})^2 &= (\overline{OB})^2\end{aligned}$$

Again, since AB is bisected at O and divided unequally at E, $\overline{AE}\cdot\overline{EB}+(\overline{OE})^2=(\overline{OB})^2$ [2.5].

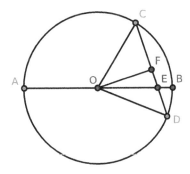

Figure 3.2.45: [3.35], case 3

It follows that

$$\begin{aligned}\overline{CE}\cdot\overline{ED}+(\overline{OE})^2 &= \overline{AE}\cdot\overline{EB}+(\overline{OE})^2 \\ \overline{CE}\cdot\overline{ED} &= \overline{AE}\cdot\overline{EB}\end{aligned}$$

4. Suppose neither chord passes through the center and they intersect at E.

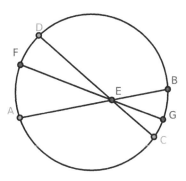

Figure 3.2.46: [3.35], case 4

Through E, construct diameter $\overline{FG}$. By case 3, the rectangle $\overline{FE}\cdot\overline{EG}=\overline{AE}\cdot\overline{EB}$ and $\overline{FE}\cdot\overline{EG}=\overline{CE}\cdot\overline{ED}$. Hence, $\overline{AE}\cdot\overline{EB}=\overline{CE}\cdot\overline{ED}$. □

Corollary. *3.35.1. If a chord of a circle is divided at any point within the circle, the rectangle contained by its segments is equal to the difference between the square of the radius and the square of the segment constructed from the center to the point of section.*

Corollary. *3.35.2. If the rectangle contained by the segments of one of two intersecting segments is equal to the rectangle contained by the segments of the other, the four endpoints are concyclic.*

Corollary. *3.35.3. If two triangles are equiangular, the rectangle contained by the non-corresponding sides about any two equal angles are equal.*

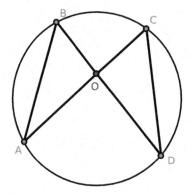

Figure 3.2.47: [3.35], Cor. 3

Proof. Let $\triangle ABO$ and $\triangle DCO$ be the equiangular triangles, and let them be placed so that the equal angles at O are vertically opposite and that the non-corresponding sides, $\overline{AO}$ and $\overline{CO}$, form segment $\overline{CA}$. Then the non-corresponding sides BO, OD form segment $\overline{BD}$. Since $\angle ABD = \angle ACD$, the points A, B, C, D are concyclic [3.21, Cor. 1]. Hence, $\overline{AO} \cdot \overline{OC} = \overline{BO} \cdot \overline{OD}$ [3.35]. □

Exercises.

1. In any triangle, the rectangle contained by two sides is equal in area to the rectangle contained by the perpendicular on the third side and the diameter of the circumscribed circle.

2. The rectangle contained by the chord of an arc and the chord of its supplement is equal to the rectangle contained by the radius and the chord of twice the supplement.

3. If the base of a triangle is given with the sum of the sides, the rectangle contained by the perpendiculars from the endpoints of the base on the external bisector of the vertical angle is given.

3.2. PROPOSITIONS FROM BOOK III   199

4. If the base and the difference of the sides is given, the rectangle contained by the perpendiculars from the endpoints of the base on the internal bisector is given.

5. Through one of the points of intersection of two circles, construct a secant so that the rectangle contained by the intercepted chords may be given, or is a maximum.

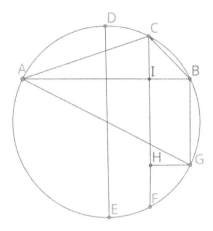

Figure 3.2.48: [3.35], #6

6. Consider the arc $AB = AF + FB$. The rectangle $\overline{AF} \cdot \overline{FB}$ is equal in area to the rectangle contained by the radius and $2 \cdot \overline{FI}$; that is, it is equal to the rectangle contained by the radius and a length equal to $\overline{CF} + \overline{BG}$. Hence, if the sum of two arcs of a circle is greater than a semicircle, the rectangle contained by their chords is equal to the rectangle contained by the radius, and the sum of the chords of the supplements of their sum and their difference.

7. Through a given point, construct a transversal cutting two given lines so that the rectangle contained by the segments intercepted between it and the line may be given.

8. If the sum of two arcs AC, CB of a circle is less than a semicircle, the rectangle $\overline{AC} \cdot \overline{CB}$ contained by their chords is equal in area to the rectangle contained by the radius and the excess of the chord of the supplement of their difference above the chord of the supplement of their sum.

Proposition 3.36. *THE AREA OF RECTANGLES CONSTRUCTED ON A TANGENT AND A POINT OUTSIDE THE CIRCLE I.*

Suppose we are given a circle and a point outside of the circle. If two segments are constructed from the point to the circle, the first of which intersects the circle at two points and the second of which is tangent to the circle, then the area of the rectangle contained by the subsegments of the first segment is equal to the square on the tangent.

Proof. Construct $\odot O$ and point P outside of $\odot O$. Then construct $\overline{PT}$ tangent to $\odot O$ at T; also construct $\overline{PA}$ such that $\overline{PA}$ intersects the circle at B and again at A. We claim that $\overline{AP} \cdot \overline{BP} = (\overline{PT})^2$. We prove this claim in two cases.

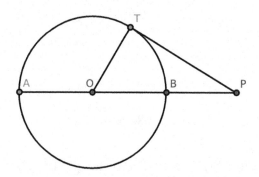

Figure 3.2.49: [3.36], case 1

Case 1: $\overline{PA}$ passes through O.

Construct $\overline{OT}$. Because $\overline{AB}$ is bisected at O and divided externally at P, the rectangle $\overline{AP} \cdot \overline{BP} + (\overline{OB})^2 = (\overline{OP})^2$ [2.6].

Since $\overline{PT}$ is a tangent to $\odot O$ and $\overline{OT}$ is constructed from the center to the point of intersection, the angle $\angle OTP$ is right [3.18]. Hence $(\overline{OT})^2 + (\overline{PT})^2 = (\overline{OP})^2$ by [1.47].

Therefore $\overline{AP} \cdot \overline{BP} + (\overline{OB})^2 = (\overline{OT})^2 + (\overline{PT})^2$. But $(\overline{OB})^2 = (\overline{OT})^2$, and so $\overline{AP} \cdot \overline{BP} = (\overline{PT})^2$.

Case 2: $\overline{PA}$ does not pass through O.

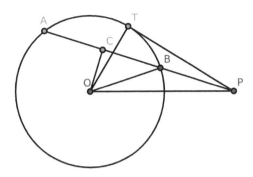

Figure 3.2.50: [3.36], case 2

Construct the perpendicular $\overline{OC} \perp \overline{AB}$; also construct $\overline{OT}$, $\overline{OB}$, $\overline{OP}$. Because $\overline{OC}$, a segment through the center, cuts $\overline{AB}$, which does not pass through the center at right angles, $\overline{OC}$ bisects $\overline{AB}$ [3.3].

Since $\overline{AB}$ is bisected at C and divided externally at P, we find that $\overline{AP} \cdot \overline{BP} + (\overline{CB})^2 = (\overline{CP})^2$ [2.6]. Adding $(\overline{OC})^2$ to each side, we obtain:

$$\begin{aligned} \overline{AP} \cdot \overline{BP} + (\overline{CB})^2 + (\overline{OC})^2 &= (\overline{CP})^2 + (\overline{OC})^2 \\ \overline{AP} \cdot \overline{BP} + (\overline{OB})^2 &= (\overline{OP})^2 \end{aligned}$$

We also have that $(\overline{OT})^2 + (\overline{PT})^2 = (\overline{OP})^2$, from which it follows that

$$\overline{AP} \cdot \overline{BP} + (\overline{OB})^2 = (\overline{OT})^2 + (\overline{PT})^2$$

Since $\overline{OB} = \overline{OT}$, $(\overline{OB})^2 = (\overline{OT})^2$, and so $\overline{AP} \cdot \overline{BP} = (\overline{PT})^2$. □

Remark. The two propositions [3.35] and [3.36] may be written as one statement: the rectangle $\overline{AP} \cdot \overline{BP}$ contained by the segments of any chord of a given circle passing through a fixed point P, either within or outside of the circle, is constant.

Proof. Suppose O is the center the circle, and construct $\overline{OA}$, $\overline{OB}$, $\overline{OP}$. Notice that $\triangle OAB$ is an isosceles triangle, and $\overline{OP}$ is a segment constructed from its vertex to a point P in the base or the extended base.

It follows that the rectangle $\overline{AP} \cdot \overline{BP}$ is equal to the difference of the squares of $\overline{OB}$ and $\overline{OP}$; therefore, it is constant. □

Corollary. *3.36.1. If two segments $\overline{AB}$, $\overline{CD}$ are extended to meet at P, and if the rectangle $\overline{AP} \cdot \overline{BP} = \overline{CP} \cdot \overline{DP}$, then the points A, B, C, D are concyclic (compare [3.35, Cor. 2]).*

Corollary. *3.36.2. Tangents to two circles from any point in their common chord are equal (compare [3.17, #6]).*

Corollary. *3.36.3. The common chords of any three intersecting circles are concurrent (compare [3.17, #7]).*

Exercises.

1. If the segment $\overline{AD}$ is constructed from the vertex A of $\triangle ABC$ which then intersects $\overline{CB}$ extended to D and creates the angle $\angle BAD = \angle ACB$, prove that $\overline{DB} \cdot \overline{DC} = (\overline{DA})^2$.

2. Prove [Cor. 3.36.1].

3.2. PROPOSITIONS FROM BOOK III

Proposition 3.37. *THE AREA OF RECTANGLES CONSTRUCTED ON A TANGENT AND A POINT OUTSIDE THE CIRCLE II.*

Suppose we are given a circle and a point outside of the circle. If two segments are constructed from the point to the circle, the first of which intersects the circle at two points, and the area of the rectangle contained by the subsegments of the first segment is equal to the square on the second segment, then the second segment is tangent to the circle.

Proof. If the rectangle ($\overline{AP} \cdot \overline{BP}$) contained by the segments of a secant and constructed from any point (P) outside of the circle ($\odot O$) is equal in area to the square on the segment ($\overline{PT}$) constructed from the same point to meet the circle, we claim that the segment which meets the circle is a tangent to that circle.

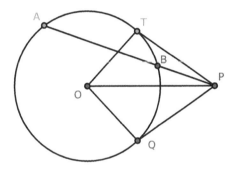

Figure 3.2.51: [3.37]

From P, construct $\overline{PQ}$ touching $\odot O$ [3.17]. Construct $\overline{OP}, \overline{OQ}, \overline{OT}$. By hypothesis, $\overline{AP} \cdot \overline{BP} = (\overline{PT})^2$; by [3.36], $\overline{AP} \cdot \overline{BP} = (\overline{PQ})^2$. Hence $(\overline{PT})^2 = (\overline{PQ})^2$, and so $\overline{PT} = \overline{PQ}$.

Consider the triangles $\triangle OTP$ and $\triangle OQP$: each have $\overline{OT} = \overline{OQ}, \overline{TP} = \overline{QP}$, and they share base $\overline{OP}$ in common. By [1.8], $\triangle OTP \cong \triangle OQP$, and so $\angle OTP = \angle OQP$.

But $\angle OQP$ is a right angle since $\overline{PQ}$ is a tangent [3.38]; hence $\angle OTP$ is right, and therefore $\overline{PT}$ is a tangent to $\odot O$ [3.16]. □

Corollary. *3.37.1. Suppose we are given a circle and a point outside of the circle where two segments are constructed from the point to the circle, the first of which intersects the circle at two points. Then the second segment is tangent to the circle if and only if the area of the rectangle contained by the subsegments of the first segment is equal to the square on the tangent.*

Exercises.

1. Construct a circle passing through two given points and fulfilling either of the following conditions:

    (a) touching a given line;

    (b) touching a given circle.

2. Construct a circle through a given point and touching two given lines; or touching a given line and a given circle.

3. Construct a circle passing through a given point having its center on a given line and touching a given circle.

4. Construct a circle through two given points and intercepting a given arc on a given circle.

5. If A, B, C, D are four collinear points and $\overline{EF}$ is a common tangent to the circles constructed upon $\overline{AB}$, $\overline{CD}$ as diameters, then prove that the triangles $\triangle AEB$, $\triangle CFD$ are equiangular.

6. The diameter of the circle inscribed in a right-angled triangle is equal to half the sum of the diameters of the circles touching the hypotenuse, the perpendicular from the right angle of the hypotenuse, and the circle constructed about the right-angled triangle.

Exam questions for chapter 3.

1. What is the subject-matter of chapter 3?

2. Define equal circles.

3. Define a chord.

4. When does a secant become a tangent?

5. What is the difference between an arc and a sector?

6. What is meant by an angle in a segment?

7. If an arc of a circle is one-sixth of the whole circumference, what is the magnitude of the angle in it?

8. What are segments?

9. What is meant by an angle standing on a segment?

10. What are concyclic points?

11. What is a cyclic quadrilateral?

12. How many intersections can a line and a circle have?

3.2. PROPOSITIONS FROM BOOK III

13. How many points of intersection can two circles have?

14. Why is it that if two circles touch they cannot have any other common point?

15. State a proposition that encompasses [3.11] and [3.12].

16. What proposition is #16 a limiting case of?

17. What is the modern definition of an angle?

18. How does the modern definition of an angle differ from Euclid's?

19. State the relations between [3.16], [3.18] and [3.19].

20. What propositions are [3.16], [3.18] and [3.19] limiting cases of?

21. How many common tangents can two circles have?

22. What is the magnitude of the rectangle of the segments of a chord constructed through a point 3.65m distant from the center of a circle whose radius is 4.25m?

23. The radii of two circles are 4.25 and 1.75 ft respectively, and the distance between their centers 6.5 ft. Find the lengths of their direct and their transverse common tangents.

24. If a point is h feet outside the circumference of a circle whose diameter is 7920 miles, prove that the length of the tangent constructed from it to the circle is $\sqrt{3h/2}$ miles.

25. Two parallel chords of a circle are 12 inches and 16 inches respectively and the distance between them is 2 inches. Find the length of the diameter.

26. What is the locus of the centers of all circles touching a given circle in a given point?

27. What is the condition that must be fulfilled that four points may be concyclic?

28. If the angle in a segment of a circle equals 1.5 right angles, what part of the whole circumference is it?

29. Mention the converse propositions of chapter 3 which are proved directly.

30. What is the locus of the midpoints of equal chords in a circle?

31. The radii of two circles are 6 and 8, and the distance between their centers is 10. Find the length of their common chord.

32. If a figure of any even number of sides is inscribed in a circle, prove that the sum of one set of alternate angles is equal to the sum of the remaining angles.

Chapter 3 exercises.

1. If two chords of a circle intersect at right angles, the sum of the squares on their segments is equal to the square on the diameter.

2. If a chord of a given circle stands opposite a right angle at a fixed point, the rectangle of the perpendiculars on it from the fixed point and from the center of the given circle is constant. Also, the sum of the squares of perpendiculars on it from two other fixed points (which may be found) is constant.

3. If through either of the points of intersection of two equal circles any line is constructed meeting them again in two points, these points are equally distant from the other intersection of the circles.

4. Construct a tangent to a given circle so that the triangle formed by it and two fixed tangents to the circle shall be:

   (a) a maximum;

   (b) a minimum.

5. If through the points of intersection A, B of two circles any two segments $\overline{ACD}$, $\overline{BEF}$ are constructed parallel to each other which meet the circles again at C, D, E, F, then we find that $\overline{CD} = \overline{EF}$.

6. In every triangle, the bisector of the greatest angle is the least of the three bisectors of the angles.

7. The circles whose diameters are the four sides of any cyclic quadrilateral intersect again in four concyclic points.

8. The four vertices of a cyclic quadrilateral determine four triangles whose orthocenters (the intersections of their perpendiculars) form an equal quadrilateral.

9. If through one of the points of intersection of two circles we construct two common chords, the segments joining the endpoints of these chords make a given angle with each other.

10. The square on the perpendicular from any point in the circumference of a circle on the chord of contact of two tangents is equal to the rectangle of the perpendiculars from the same point on the tangents.

11. Find a point on the circumference of a given circle such that the sum of the squares on whose distances from two given points is either a maximum or a minimum.

12. Four circles are constructed on the sides of a quadrilateral as diameters. Prove that the common chord of any two on adjacent sides is parallel to the common chord of the remaining two.

13. The rectangle contained by the perpendiculars from any point in a circle on the diagonals of an inscribed quadrilateral is equal to the rectangle contained by the perpendiculars from the same point on either pair of opposite sides.

14. The rectangle contained by the sides of a triangle is greater than the square on the internal bisector of the vertical angle by the rectangle contained by the segments of the base.

15. If through A, one of the points of intersection of two circles, we construct any line $\overleftrightarrow{ABC}$ which cuts the circles again at B and C, the tangents at B and C intersect at a given angle.

16. If a chord of a given circle passes through a given point, the locus of the intersection of tangents at its endpoints is a straight line.

17. The rectangle contained by the distances of the point where the internal bisector of the vertical angle meets the base and the point where the perpendicular from the vertex meets it from the midpoint of the base is equal to the square on half the difference of the sides.

18. State and prove the proposition analogous to [3.17] for the external bisector of the vertical angle.

19. The square on the external diagonal of a cyclic quadrilateral is equal to the sum of the squares on the tangents from its endpoints to the circumscribed circle.

20. If a "movable" circle touches a given circle and a given line, the chord of contact passes through a given point.

21. If A, B, C are three points in the circumference of a circle, and D, E are the midpoints of the arcs AB, AC, and if the segment $\overline{DE}$ intersects the chords $\overline{AB}$, $\overline{AC}$ at F and G, then $\overline{AF} = \overline{AG}$.

22. If a cyclic quadrilateral is such that a circle can be inscribed in it, the lines joining the points of contact are perpendicular to each other.

23. If through the point of intersection of the diagonals of a cyclic quadrilateral the minimum chord is constructed, that point will bisect the part of the chord between the opposite sides of the quadrilateral.

24. Given the base of a triangle, the vertical angle, and either the internal or the external bisector at the vertical angle, construct the triangle.

25. If through the midpoint A of a given arc BAC we construct any chord $\overline{AD}$, cutting BC at E, then the rectangle $\overline{AD} \cdot \overline{AE}$ is constant.

26. The four circles circumscribing the four triangles formed by any four lines pass through a common point.

27. If X, Y, Z are any three points on the three sides of a triangle $\triangle ABC$, the three circles about the triangles $\triangle YAZ$, $\triangle ZBX$, $\triangle XCY$ pass through a common point.

28. If the position of the common point in the previous exercise are given, the three angles of the triangle $\triangle XYZ$ are given, and conversely.

29. Place a given triangle so that its three sides shall pass through three given points.

30. Place a given triangle so that its three vertices shall lie on three given lines.

31. Construct the largest triangle equiangular to a given one whose sides shall pass through three given points.

32. Construct the smallest possible triangle equiangular to a given one whose vertices shall lie on three given lines.

33. Construct the largest possible triangle equiangular to a given triangle whose sides shall touch three given circles.

34. If two sides of a given triangle pass through fixed points, the third touches a fixed circle.

35. If two sides of a given triangle touch fixed circles, the third touches a fixed circle.

36. Construct an equilateral triangle having its vertex at a given point and the endpoints of its base on a given circle.

37. Construct an equilateral triangle having its vertex at a given point and the endpoints of its base on two given circles.

38. Place a given triangle so that its three sides touch three given circles.

39. Circumscribe a square about a given quadrilateral.

40. Inscribe a square in a given quadrilateral.

41. Construct the following circles:

   (a) orthogonal (cutting at right angles) to a given circle and passing through two given points;

   (b) orthogonal to two others, and passing through a given point;

   (c) orthogonal to three others.

42. If from the endpoints of a diameter $\overline{AB}$ of a semicircle two chords $\overline{AD}$, $\overline{BE}$ are constructed which meet at C, we find that $\overline{AC} \cdot \overline{AD} + \overline{BC} \cdot \overline{BE} = (\overline{AB})^2$.

43. If $ABCD$ is a cyclic quadrilateral, and if we construct any circle passing through the points A and B, another through B and C, a third through C and D, and a fourth through D and A, then these circles intersect successively at four other points E, F, G, H, forming another cyclic quadrilateral.

44. If $\triangle ABC$ is an equilateral triangle, what is the locus of the point M, if $\overline{MA} = \overline{MB} + \overline{MC}$?

45. In a triangle, given the sum or the difference of two sides and the angle formed by these sides both in magnitude and position, the locus of the center of the circumscribed circle is a straight line.

46. Construct a circle:

   (a) through two given points which bisect the circumference of a given circle;

3.2. PROPOSITIONS FROM BOOK III

(b) through one given point which bisects the circumference of two given circles.

47. Find the locus of the center of a circle which bisects the circumferences of two given circles.

48. Construct a circle which bisects the circumferences of three given circles.

49. If $\overline{CD}$ is a perpendicular from any point C in a semicircle on the diameter $\overline{AB}$, $\bigcirc EFG$ is a circle touching $\overline{DB}$ at E, $\overline{CD}$ at F, and the semicircle at G, then prove that:

(a) the points A, F, G are collinear;

(b) $\overline{AC} = \overline{AE}$.

50. Being given an obtuse-angled triangle, construct from the obtuse angle to the opposite side a segment whose square is equal to the rectangle contained by the segments into which it divides the opposite side.

51. If O is a point outside a circle whose center is E and two perpendicular segments passing through O intercept chords $\overline{AB}$, $\overline{CD}$ on the circle, then prove that $(\overline{AB})^2 + (\overline{CD})^2 + 4 \cdot (\overline{OE})^2 = 8 \cdot (\overline{R})^2$.

52. The sum of the squares on the sides of a triangle is equal to twice the sum of the rectangles contained by each perpendicular and the portion of it comprised between the corresponding vertex and the orthocenter. It is also equal to $12 \cdot (\overline{R})^2$ minus the sum of the squares of the distances of the orthocenter from the vertices.

53. If two circles touch at C, if D is any point outside the circles at which their radii through C stands opposite equal angles, and if DE, DF are tangent from D, prove that $\overline{DE} \cdot \overline{DF} = (\overline{CD})^2$.

Chapter 4

Inscription and Circumscription

This chapter contains sixteen propositions: four relate to triangles, four to squares, four to pentagons, and four to other figures.

4.1 Definitions

1. If two polygons are related such that the vertices of one lie on the sides of the other, then:

   (a) the inner figure is said to be *inscribed* in the outer figure;

   (b) the outer figure is said to be *circumscribed* around or about the inner figure.

2. A polygon is said to be *inscribed* in a circle when all of its vertices intersect the circumference. Reciprocally, a polygon is said to be *circumscribed* about or around a circle when each of its sides touch the circle.

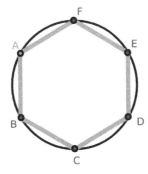

Figure 4.1.1: The hexagon is inscribed in the circle, and the circle is circumscribed about the hexagon.

3. A circle is said to be *inscribed* in a polygon when it touches each side of the figure. Reciprocally, a circle is said to be *circumscribed* about or around a polygon when it passes through each vertex of the figure.

4. A polygon which is both equilateral and equiangular is said to be *regular*.

5. The bisectors of the three internal angles of a triangle are concurrent. Their point of intersection is called the *incenter* of the triangle.

6. The circle from [4.5] is called the *circumcircle*, its radius the *circumradius*, and its center the *circumcenter* of the triangle.

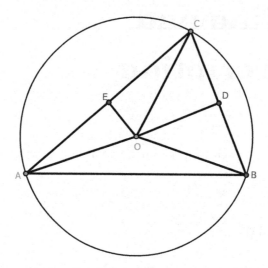

Figure 4.1.2: [4.5]

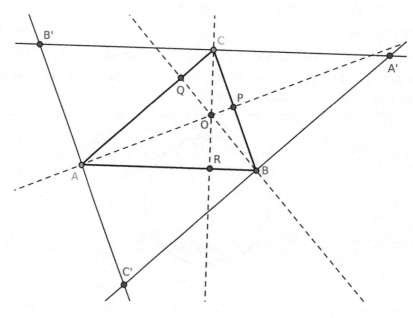

Figure 4.1.3: [4.5, #2]

4.1. DEFINITIONS

7. In [4.5, #2], construct $\circ O$ such that its radius equals $\overline{OA} \cdot \overline{OP} = \overline{OB} \cdot \overline{OQ} = \overline{OC} \cdot \overline{OR}$; this circle is defines as the *polar circle* of the triangle $\triangle ABC$.

8. The nine-points circle is a circle that can be constructed for any given triangle. It is so named because it passes through nine significant concyclic points defined from the triangle. These nine points are:

(a) the midpoint of each side of the triangle

(b) the foot of each altitude

(c) the midpoint of the line segment from each vertex of the triangle to the orthocenter (where the three altitudes meet; these line segments lie on their respective altitudes).[1]

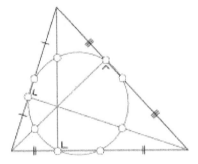

Figure 4.1.4: [4.5, #4] The nine-points circle

---

[1] https://en.wikipedia.org/wiki/Nine-point_circle

4.2 Propositions from Book IV

Proposition 4.1. *CONSTRUCTING A CHORD INSIDE A CIRCLE.*

Construct an arbitrary circle and an arbitrary segment such that the segment is less than or equal to the length of the diameter of the circle. It is possible to construct a chord within the circle equal to the length of the segment.

Proof. Construct $\circ ABC$ with diameter $\overline{AC}$ and the segment $\overline{DG} \leq \overline{AC}$. We wish to construct a chord in $\circ ABC$ equal in length to $\overline{DG}$.

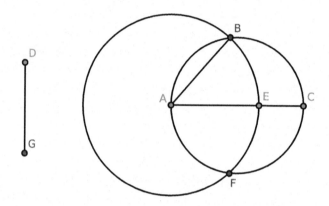

Figure 4.2.1: [4.1]

If $\overline{DG} = \overline{AC}$, then the required chord already exists (the diameter of $\circ A$).

If $\overline{DG} < \overline{AC}$, cut sub-segment $\overline{AE}$ from diameter $\overline{AC}$ such that $\overline{AE} = \overline{DG}$ [1.3]. With A as center and $\overline{AE}$ as radius, construct the circle $\circ A$, cutting the circle $\circ ABC$ at the points B and F.

Construct $\overline{AB}$: we claim that $\overline{AB}$ is the required chord.

Notice that $\overline{AB} = \overline{AE}$. Since $\overline{AE} = \overline{DG}$ by construction, $\overline{AB} = \overline{DG}$. Since $\overline{AB}$ is a chord of $\circ ABC$, the construction is complete. □

4.2. PROPOSITIONS FROM BOOK IV

Proposition 4.2. *INSCRIBING A TRIANGLE INSIDE A CIRCLE.*

In a given circle, it is possible to inscribe a triangle that is equiangular to a given triangle.

Proof. We wish to inscribe a triangle equiangular to $\triangle DEF$ in $\bigcirc ABC$.

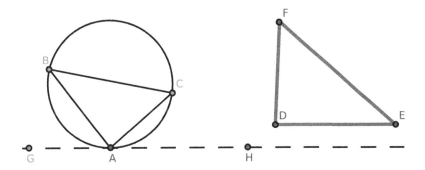

Figure 4.2.2: [4.2]

At A on the circumference of $\bigcirc ABC$, construct the tangent line $\overleftrightarrow{GAH}$. Construct $\angle HAC = \angle DEF$, $\angle GAB = \angle DFE$ [1.23], and segment $\overline{BC}$. We claim that $\triangle ABC$ fulfills the required conditions.

Since $\angle DEF = \angle HAC$ by construction and $\angle HAC = \angle ABC$ by [3.32], $\angle DEF = \angle ABC$. Similarly, $\angle DFE = \angle ACB$. By [1.32], $\angle FDE = \angle BAC$.

This completes the construction. □

Proposition 4.3. *CIRCUMSCRIBING A TRIANGLE ABOUT A CIRCLE.*

It is possible to circumscribe a triangle about a circle such that the triangle is equiangular to a given triangle.

Proof. We wish to to construct a triangle equiangular to $\triangle DEF$ about $\odot O$.

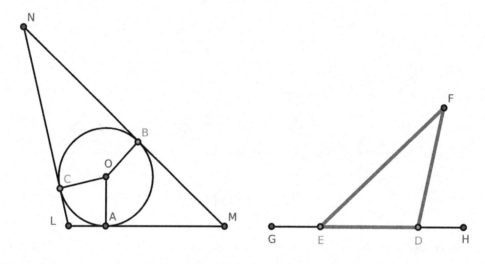

Figure 4.2.3: [4.3]

Extend side $\overline{DE}$ of $\triangle DEF$ to $\overline{GH}$, and construct $\overline{OA}$. Construct $\angle AOB = \angle GEF$ and $\angle AOC = \angle HDF$ [1.23]. At the points A, B, and C, construct the tangents $\overline{LM}$, $\overline{MN}$, and $\overline{NL}$ to $\odot O$. We claim that $\triangle LMN$ fulfills the required conditions.

Because $\overline{AM}$ touches $\odot O$ at A, $\angle OAM$ is right [3.18]. Similarly, $\angle MBO$ is right; but the sum of the four angles of the quadrilateral $OAMB$ is equal to four right angles [1.32, Cor. 3]. Therefore the sum of the two remaining angles $\angle AOB + \angle AMB$ equals two right angles.

By [1.13], $\angle GEF + \angle FED =$ two right angles, and so $\angle AOB + \angle AMB = \angle GEF + \angle FED$. But $\angle AOB = \angle GEF$ by construction; hence $\angle AMB = \angle FED$. Similarly, $\angle ALC = \angle EDF$. By [1.32], $\angle BNC = \angle DFE$, and so $\triangle LMN$ is equiangular to $\triangle DEF$. This completes the construction. □

Proposition 4.4. *INSCRIBING A CIRCLE IN A TRIANGLE.*

It is possible to inscribe a circle in a given triangle.

Proof. We wish to inscribe $\odot O$ in $\triangle ABC$.

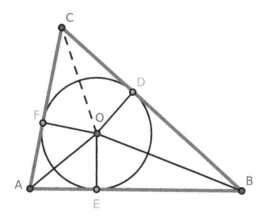

Figure 4.2.4: [4.4]

Bisect angles $\angle CAB$ and $\angle ABC$ of $\triangle ABC$ with $\overline{AO}$ and $\overline{BO}$, respectively. We claim that O, their point of intersection, is the center of the required circle.

From O construct $\overline{OD} \perp \overline{CB}$, $\overline{OE} \perp \overline{AB}$, and $\overline{OF} \perp \overline{AC}$. Consider the triangles $\triangle OAE$ and $\triangle OAF$: $\angle OAE = \angle OAF$ by construction; $\angle AEO = \angle AFO$ since each is right; each triangle shares side $\overline{OA}$. By [1.26], $\triangle OAE \cong \triangle OAF$, and so $\overline{OE} = \overline{OF}$.

Similarly, $\overline{OD} = \overline{OF}$, and so $\overline{OD} = \overline{OE} = \overline{OF}$. As a consequence of [3.9], the circle constructed with O as center and $\overline{OD}$ as radius will intersect the points D, E, F.

Since each of the angles $\angle ODB$, $\angle OEA$, $\angle OFA$ is right, each segment touches the respective sides of the triangle $\triangle ABC$ [3.16]. Therefore, the circle $\odot O$ is inscribed in the triangle $\triangle ABC$, which completes the construction. □

Exercises.

1. In [4.4]: if $\overline{OC}$ is constructed, prove that $\angle ACB$ is bisected. Hence, the existence of the *incenter* of a triangle is proven. [See the final chapter for a solution.]

2. If the sides BC, CA, AB of the triangle $\triangle ABC$ are written as a, b, c, and half the sum of their side-lengths is defined as s, prove that the distances of the vertices A, B, and C of the triangle from the points of contact of the inscribed circle are respectively $s - a, s - b, s - c$.

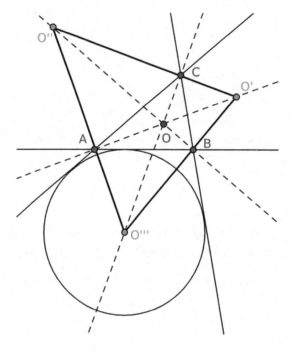

Figure 4.2.5: [4.4, #3]

3. If the external angles of the triangle $\triangle ABC$ are bisected as in the above Figure, prove that the three vertices O', O'', O''' of the triangle formed by the three bisectors are the centers of three circles, each touching one side externally and the other two when extended. These three circles are defined as the *escribed circles* of the triangle $\triangle ABC$.

4. Prove that center of the inscribed circle, the center of each escribed circle, and two of the vertices of the triangle are concyclic. Also, prove that any two of the escribed centers are concyclic with the corresponding two of the vertices of the triangle.

5. Of the four points O, O', O'', O''', prove that any one is the orthocenter of the triangle formed by the remaining three.

6. In the above figure, prove that $\triangle BCO'$, $\triangle CAO''$, and $\triangle ABO'''$ are equiangular.

7. Given the base of a triangle, the vertical angle, and the radius of the inscribed or any of the escribed circles, construct it.

4.2. PROPOSITIONS FROM BOOK IV

Proposition 4.5. *CIRCUMSCRIBING A CIRCLE ABOUT A TRIANGLE.*

It is possible to circumscribe a circle about a given triangle.

Proof. We wish to construct $\odot O$ about $\triangle ABC$.

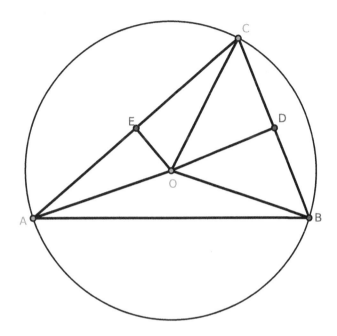

Figure 4.2.6: [4.5]

Bisect sides AC and BC of $\triangle ABC$ at the points E and D, respectively. Construct $\overline{DO} \perp \overline{BC}$ and $\overline{EO} \perp \overline{CA}$. We claim that O, the point of intersection of the perpendicular segments, is the center of the required circle.

Construct $\overline{OA}, \overline{OB}, \overline{OC}$, and consider triangles $\triangle BDO$ and $\triangle CDO$: sides $BD = CD$ by construction, the triangles share side DO in common, and $\angle BDO = \angle CDO$ because each is right. By [1.4], $\triangle BDO \cong \triangle CDO$, and so $\overline{BO} = \overline{OC}$.

Similarly, $\overline{AO} = \overline{OC}$, and so $\overline{AO} = \overline{BO} = \overline{CO}$. As a consequence of [3.9], a circle can be constructed with O as its center and $\overline{OA}$ as its radius such that the circumference $\odot O$ will pass through A, B, and C. Thus $\odot O$ is circumscribed about the triangle $\triangle ABC$. □

Corollary. *4.5.1. Since the perpendicular from O to $\overline{AB}$ bisects $\overline{AB}$ by [3.3], we see that the perpendiculars at the midpoints of the sides of a triangle are concurrent. (See also [Def. 4.7].)*

Exercises.

1. Prove that the three altitudes of a triangle ($\triangle ABC$) are concurrent. (This proves the existence of the *orthocenter* of a circle.) [See the final chapter for a solution.]

2. In the figure below, prove that the three rectangles $\overline{OA} \cdot \overline{OP}, \overline{OB} \cdot \overline{OQ}, \overline{OC} \cdot \overline{OR}$ are equal in area. (See also [Def. 4.7].)

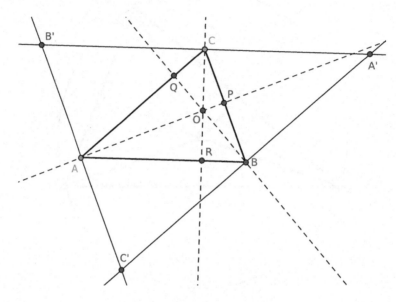

Figure 4.2.7: [4.5, #2]

3. If the altitudes of a triangle are extended to meet a circumscribed circle, prove that the intercepts between the orthocenter and the circle are bisected by the sides of the triangle.

4. Prove that the circumcircle of a triangle is the "nine points circle" of each of the four triangles formed by joining the centers of the inscribed and escribed circles. (See [Def. 4.8].)

5. Prove that the radius of the "nine points circle" of a triangle is equal to half its circumradius. (See [Def. 4.8].)

6. Prove that the distances between the vertices of a triangle and its orthocenter are respectively the doubles of the perpendiculars from the circumcenter on the sides.

Remark. The orthocenter, centroid, and circumcenter of any triangle are collinear; they lie on the Euler line[2].

---

[2]https://en.wikipedia.org/wiki/Euler_line

4.2. PROPOSITIONS FROM BOOK IV

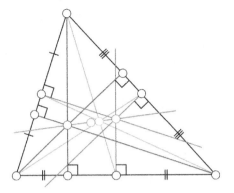

Figure 4.2.8: The Euler Line

Proposition 4.6. *INSCRIBING A SQUARE IN A CIRCLE.*

It is possible to to inscribe a square in a given circle.

Proof. We wish to inscribe the square $\square ABCD$ in $\bigcirc O$.

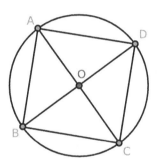

Figure 4.2.9: [4.6]

Construct diameters $\overline{AC}$ and $\overline{BD}$ such that $\overline{AC} \perp \overline{BD}$. Also construct $\overline{AB}$, $\overline{BC}$, $\overline{CD}$, and $\overline{DA}$. We claim that $\square ABCD$ is the required square.

Notice that the four angles at O are equal since they are right angles. Hence the arcs on which they stand are equal [3.26] and the four chords on which they stand are equal in length [3.29]. Therefore the figure $\square ABCD$ is equilateral.

Again, since $\overline{AC}$ is a diameter, the angle $\angle ABC$ is right [3.31]. Similarly, the remaining angles are right. It follows that $\square ABCD$ is a square inscribed in $\bigcirc O$, which completes the construction. □

Proposition 4.7. *CIRCUMSCRIBING A SQUARE ABOUT A CIRCLE.*

It is possible to circumscribe a square about a given circle.

Proof. We wish to construct the square $\square EHGF$ about $\circ O$.

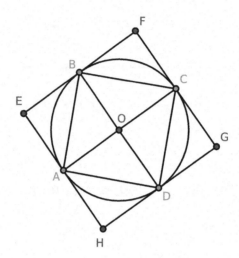

Figure 4.2.10: [4.7]

Through O construct diameters $\overline{AC}$ and $\overline{BD}$ such that $\overline{AC} \perp \overline{BD}$, and at the points A, B, C, and D construct the tangential segments $\overline{HE} = \overline{EF} = \overline{FG} = \overline{GH}$. We claim that $\square EFGH$ is the required square.

Since $\overline{AE}$ touches the circle at A, the angle $\angle EAO$ is right [3.18] and therefore equal to $\angle BOC$, which is right by construction. Thus $\overline{AE} \parallel \overline{OB}$ and $\overline{EB} \parallel \overline{AO}$.

Since $\overline{AO} = \overline{OB}$ (both are radii of $\circ O$), the figure $HDOA$ is a rhombus. Since the angle $\angle AOB$ is right, $\square AOBE$ is a square.

Similarly, each of the figures $\square OCFB$, $\square DGCO$, and $\square HDOA$ is a square. Similarly to the above, $\square EHGF$ is also a square circumscribed about $\circ O$, which completes the construction. $\square$

Corollary. *4.7.1. The circumscribed square, $\square EHGF$, has double the area of the inscribed square, $\square BCDA$.*

Exercises.

1. Prove [Cor. 4.7.1]. [See the final chapter for a solution.]

4.2. PROPOSITIONS FROM BOOK IV

Proposition 4.8. *INSCRIBING A CIRCLE IN A SQUARE.*

It is possible to inscribe a circle in a given square.

Proof. We wish to inscribe $\odot O$ in $\square EHGF$.

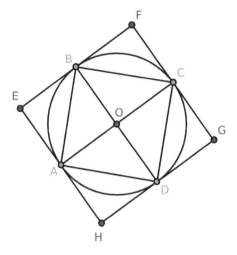

Figure 4.2.11: [4.8]

Bisect adjacent sides $\overline{EH}$ and $\overline{EF}$ at A and B, respectively. Through A and B, construct $\overline{AC} \perp \overline{EH}$ and $\overline{BD} \perp \overline{EF}$. We claim that O, the point of intersection of these parallel segments, is the center of the required circle, $\odot O$.

Because $\square EAOB$ is a parallelogram, its opposite sides are equal; therefore $\overline{OA} = \overline{EB}$. But $\overline{EB}$ is half the side of $\square EHGF$, and so $\overline{OA}$ = half of the side of $\square EHGF$. This is also true for each of the segments $\overline{OB}$, $\overline{OC}$, and $\overline{OD}$, *mutatis mutandis*. Hence

$$\overline{OA} = \overline{OB} = \overline{OC} = \overline{OD}$$

As a consequence of [3.9], O is the center of $\odot O$. And since these segments are perpendicular to the sides of the given square, the circle constructed with O as center and $\overline{OA}$ as radius is inscribed in the square. This completes the construction. $\square$

Proposition 4.9. *CIRCUMSCRIBING A CIRCLE ABOUT A GIVEN SQUARE.*

It is possible to circumscribe a circle about a given square.

Proof. We wish to construct $\odot O$ about $\square ABCD$.

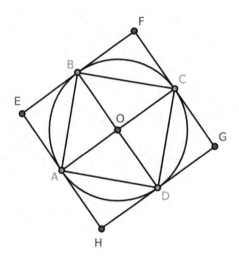

Figure 4.2.12: [4.9]

Construct perpendicular diagonals $\overline{AC}$ and $\overline{BD}$ intersecting at O. We claim that O is the center of the required circle.

Consider $\triangle ABD$ and $\triangle CBD$: since $\square ABCD$ is a square, $\overline{DA} = \overline{AB} = \overline{BC} = \overline{CD}$. The triangles share side $\overline{BD}$. By [1.8], $\triangle ABD \cong \triangle CBD$, and so $\angle ABD = \angle CBD$. Since $\angle ABC = \angle ABD + \angle CBD$, $\angle ABC$ is bisected by $\overline{BD}$. Similarly, we can prove that $\angle ADC$ is bisected by $\overline{BD}$ and that $\angle DAB$ and $\angle BCD$ are bisected by $\overline{AC}$.

Since $\square ABCD$ is a square, $\angle ABC = \angle BCD = \angle CDA = \angle DAB$, and so $\angle ABO = \angle CBO = \angle BCO = \angle DCO = \angle CDO = \angle ADO = \angle DAO = \angle BAO$.

Consider $\triangle ABO$ and $\triangle CBO$: $\angle ABO = \angle CBO$ by the above, they share side $\overline{OB}$, and $\overline{AB} = \overline{BC}$ by the above. By [1.4], $\triangle ABO \cong \triangle CBO$. Also notice that $\angle BAO = \angle ABO = \angle CBO = \angle BCO$, and so each triangle is isosceles. By [1.6], $\overline{AO} = \overline{BO} = \overline{CO}$.

As a consequence of [3.9], O is the center of $\odot O$ with radius $= \overline{OA}$ which intersects B, C, and D and is clearly constructed about the square $\square ABCD$. This completes the construction. □

4.2. PROPOSITIONS FROM BOOK IV

Proposition 4.10. *CONSTRUCTION OF AN ISOSCELES TRIANGLE WITH BASE ANGLES DOUBLE THE VERTICAL ANGLE.*

It is possible to construct an isosceles triangle such that each base angle is double the vertical angle.

Proof. We wish to construct an isosceles triangle $\triangle ABD$ such that

$$\angle DBA = 2 \cdot \angle DAB = \angle BDA$$

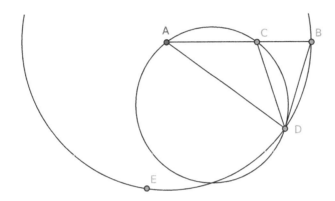

Figure 4.2.13: [4.10]

Construct $\overline{AB}$ and divide it at C such that $\overline{AB} \cdot \overline{BC} = \left(\overline{AC}\right)^2$ [2.6].

With A as center and $\overline{AB}$ as radius, construct $\circ A$; on its circumference construct $\overline{BD} = \overline{AC}$ [4.1]. Also construct $\overline{AD}$. We claim that $\triangle ABD$ fulfills the required conditions.

Construct $\overline{CD}$ and the circle $\circ ACD$ about $\triangle ACD$ [4.5].

Since $\overline{BD} = \overline{AC}$, $\overline{AB} \cdot \overline{BC} = \left(\overline{BD}\right)^2$. Notice that B is outside of $\circ ACD$. By [3.37] $\overline{BD}$ is tangent to $\circ ACD$. By [3.32], $\angle BDC = \angle DAC$, and so

$$\angle CDA + \angle BDC = \angle BDA = \angle CDA + \angle DAC$$

By [1.32], $\angle BCD = \angle CDA + \angle DAC$, and so $\angle BDA = \angle BCD$.

Since $\overline{AB} = \overline{AD}$, by [1.5] $\angle BDA = \angle BCD = \angle CBD$.

Since $\angle CBD = \angle BCD$, by [1.6] $\overline{BD} = \overline{DC}$. Hence $\overline{BD} = \overline{DC} = \overline{AC}$.

Again from [1.5], $\angle CDA = \angle DAC$, and so $\angle CDA + \angle DAC = 2 \cdot \angle DAC$.

By the above, $\angle BCD = \angle CDA + \angle DAC$, and so $\angle BCD = 2 \cdot \angle DAC = 2 \cdot \angle DAB$.

Since $\angle BCD = \angle CBD = \angle DBA$, $\angle DBA = 2 \cdot \angle DAB$.

Since $\angle BDA = \angle BCD$, $\angle DBA = 2 \cdot \angle DAB = \angle BDA$, which completes the construction.

□

Exercises.

1. Prove that $\triangle ACD$ is an isosceles triangle whose vertical angle is equal to three times each of the base angles. [See the final chapter for a solution.]

2. Prove that $\overline{BD}$ is the side of a regular decagon inscribed in the circle $\bigcirc BDE$.

3. If $\overline{DB}$, $\overline{DE}$, and $\overline{EF}$ are consecutive sides of a regular decagon inscribed in a circle, prove that $\overline{BF} - \overline{BD} =$ radius of a circle.

4. If E is the second point of intersection of the circle $\bigcirc ACD$ with $\bigcirc BDE$, prove that $\overline{DE} = \overline{DB}$. If $\overline{AE}, \overline{BE}, \overline{CE}$, and $\overline{DE}$ are constructed, then triangles $\triangle ACE$ and $\triangle ADE$ are each congruent with $\triangle ABD$.

5. Prove that $\overline{AC}$ is the side of a regular pentagon inscribed in the circle $\bigcirc ACD$, and $\overline{EB}$ the side of a regular pentagon inscribed in the circle $\bigcirc BDE$.

6. Since $\triangle ACE$ is an isosceles triangle, $(\overline{EB})^2 - (\overline{EA})^2 = \overline{AB} \cdot \overline{BC} = (\overline{BD})^2$; that is, prove that the square of the side of a pentagon inscribed in a circle exceeds the square of the side of the decagon inscribed in the same circle by the square of the radius.

4.2. PROPOSITIONS FROM BOOK IV

Proposition 4.11. *INSCRIBING A REGULAR PENTAGON IN A GIVEN CIRCLE.*

It is possible to inscribe a regular pentagon in a given circle.

Proof. We wish to inscribe a regular pentagon in $\odot ABC$.

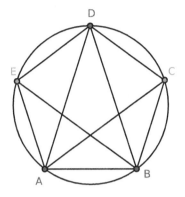

Figure 4.2.14: [4.11]

Construct any isosceles triangle having each base angle equal to double the vertical angle [4.10], and then construct $\triangle ABD$ equiangular to that triangle such that it is inscribed in $\odot ABC$ [4.4].

Bisect the angles $\angle DAB$ and $\angle ABD$ by constructing $\overline{AC}$ and $\overline{BE}$, respectively. Also construct $\overline{EA}$, $\overline{ED}$, $\overline{DC}$, and $\overline{CB}$. We claim that the figure $ABCDE$ is a regular pentagon.

Since $\angle DAB = \angle ABD = 2 \cdot \angle ADB$ by construction, $\overline{AC}$ bisects $\angle DAB$, and $\overline{BE}$ bisects $\angle ABD$, and so
$$\angle BAC = \angle CAD = \angle ADB = \angle DBE = \angle EBA$$

By [Cor. 3.29.1], the chords on which these angles stand are equal in length:

$$\overline{AB} = \overline{BC} = \overline{CD} = \overline{DE} = \overline{EA}$$

Hence $ABCDE$ is equilateral.

Again, because the arcs AB and DE are equal in length, if we add the arc BCD to both, then the arc $ABCD$ is equal in length to the arc $BCDE$, and therefore the angles $\angle AED$, $\angle BAE$ which stand on them are equal [3.27].

Similarly, it can be shown that all of these angles are equal; therefore $ABCDE$ is equiangular and a regular pentagon; this proves our claim. □

Exercises.

1. Prove that the figure formed by the five diagonals of a regular pentagon is another regular pentagon.

2. If the alternate sides of a regular pentagon are extended to intersect, the five points of meeting form another regular pentagon.

3. Prove that every two consecutive diagonals of a regular pentagon divide each other in the extreme and mean ratio [2.11].

4. Being given a side of a regular pentagon, construct it.

5. Divide a right angle into five equal parts.

4.2. PROPOSITIONS FROM BOOK IV

Proposition 4.12. *CIRCUMSCRIBING A REGULAR PENTAGON ABOUT A GIVEN CIRCLE.*

It is possible to circumscribe a regular pentagon about a given circle.

Proof. We wish to construct a regular pentagon about $\odot O$.

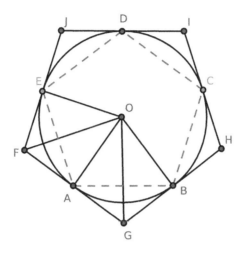

Figure 4.2.15: [4.12]

Use [4.11] to inscribe a regular pentagon inside $\odot O$ with vertices at points A, B, C, D, and E; at these points, construct equal tangential segments $\overline{FG}, \overline{GH}, \overline{HI}, \overline{IJ}$, and $\overline{JF}$. We claim that $FGHIJ$ is the required circumscribed regular pentagon.

Construct $\overline{OE}, \overline{OA}$, and $\overline{OB}$. Because the angles $\angle OAF$ and $\angle OEF$ of the quadrilateral $AOEF$ are right angles [3.18], the sum of the two remaining angles $\angle AOE + \angle AFE$ equals two right angles. Similarly, the sum $\angle AOB + \angle AGB$ equals two right angles; hence $\angle AOE + \angle AFE = \angle AOB + \angle AGB$. But $\angle AOE = \angle AOB$ because they stand on equal segments $\overline{AE}$ and $\overline{AB}$ [3.27]. Hence $\angle AFE = \angle AGB$. Similarly, the remaining angles of the figure $FGHIJ$ are equal, and so $FGHIJ$ is equiangular.

Now construct $\overline{OF}$ and $\overline{OG}$ and consider $\triangle EOF$ and $\triangle AOF$: $\overline{AF} = \overline{FE}$ [3.17, #1], the triangles share side $\overline{FO}$, and $\overline{AO} = \overline{EO}$ since each are radii of $\odot O$. By [1.8], $\triangle EOF \cong \triangle AOF$, and so $\angle AFO = \angle EFO$; or, $\angle AFE$ is bisected at F.

Since $\angle AFE = \angle AFO + \angle EFO = 2 \cdot \angle AFO$, it follows that $\angle AFO = \frac{1}{2}\angle AFE$. Similarly, $\angle AGO = \frac{1}{2}\angle AGB$.

Consider $\triangle AFO$ and $\triangle AGO$: $\angle AFE = \angle AGB$ implies that $\angle AFO = \angle AGO$; $\angle FAO = \angle GAO$ since each are right angles; finally, each shares side $\overline{AO}$. By [1.26], $\triangle AFO \cong \triangle AGO$, and so $\overline{AF} = \overline{AG}$.

As a consequence, $\overline{GF} = 2 \cdot \overline{AF}$; similarly, $\overline{JF} = 2 \cdot \overline{EF}$. And since $\overline{AF} = \overline{EF}, \overline{GF} = \overline{JF}$, and so on for all remaining sides. Therefore, $FGHIJ$ is equilateral and equiangular; thus, it is a regular pentagon, which proves our claim. □

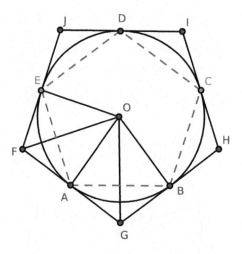

Remark. This proposition is a particular case of the following general theorem (which has an analogous proof): "If tangents are constructed on a circle at the vertices of an inscribed regular polygon with a finite number of sides, they will form a regular polygon with the same number of sides circumscribed to the circle."

4.2. PROPOSITIONS FROM BOOK IV

Proposition 4.13. *INSCRIBING A CIRCLE IN A REGULAR PENTAGON.*

It is possible to inscribe a circle in a regular pentagon.

Proof. We wish to inscribe $\odot O$ in regular pentagon $ABCDE$.

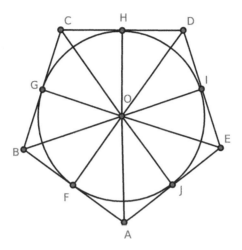

Figure 4.2.16: [4.13]

Bisect two adjacent angles $\angle JAF$ and $\angle FBG$ by constructing $\overline{AO}$ and $\overline{BO}$, respectively; we claim that the point of intersection of the bisectors, O, is the center of the required circle.

Construct $\overline{CO}$ as well as perpendiculars from O to the five sides of the pentagon.

Consider $\triangle ABO$ and $\triangle CBO$: $\overline{AB} = \overline{BC}$ by hypothesis, $\angle ABO = \angle CBO$ by construction, and each shares side $\overline{BO}$. By [1.4] $\triangle ABO \cong \triangle CBO$, and so $\angle BAO = \angle BCO$; however, $\angle BAO = \frac{1}{2} \cdot \angle BAE$ by construction. Therefore

$$\angle BCO = \frac{1}{2} \cdot \angle BAE = \frac{1}{2} \cdot \angle BCD$$

Hence $\overline{CO}$ bisects the angle $\angle BCD$. Similarly, we can show that $\overline{DO}$ bisects $\angle HDI$ and that $\overline{EO}$ bisects $\angle IEJ$.

Consider $\triangle BOF$ and $\triangle BOG$: $\angle OFB = \angle OGB$ since each are right, $\angle OBF = \angle OBG$ because $\overline{OB}$ bisects $\angle ABC$ by construction, and each shares side $\overline{OB}$. By [1.26], $\triangle BOF \simeq \triangle BOG$, and so $\overline{OF} = \overline{OG}$.

Similarly, all the perpendiculars from O to the sides of the pentagon are equal. By [3.9], the circle whose center is O with radius $\overline{OF}$ is inscribed in regular pentagon $ABCDE$, which completes the construction. □

Proposition 4.14. *CIRCUMSCRIBING A CIRCLE ABOUT A REGULAR PENTAGON.*

It is possible to circumscribe a circle about a regular pentagon.

Proof. We wish to construct $\odot O$ about regular pentagon $ABCDE$.

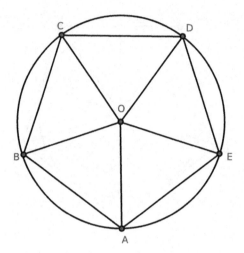

Figure 4.2.17: [4.14]

Bisect adjacent angles $\angle BAE$ by $\overline{AO}$ and $\angle ABC$ by $\overline{BO}$. We claim that O, the point of intersection of the bisectors, is the center of the required circle.

Similarly, construct $\overline{OC}$, $\overline{OD}$, and $\overline{OE}$. Consider $\triangle ABO$ and $\triangle CBO$: $\overline{AB} = \overline{BC}$ and $\angle ABO = \angle CBO$ by construction, and the triangles share $\overline{BO}$ in common. By [1.4], $\triangle ABO \cong \triangle CBO$, and so $\angle BAO = \angle BCO$.

But $\angle BAE = \angle BCD$ since $ABCDE$ is a regular pentagon. Since $\angle BAO = \frac{1}{2}\angle BAE$ by construction, $\angle BCO = \frac{1}{2}\angle BCD$; hence, $\overline{CO}$ bisects $\angle BCD$. Similarly, it can be shown that $\overline{DO}$ bisects $\angle CDE$ and $\overline{EO}$ bisects $\angle DEA$.

Because $\angle EAB = \angle ABC$, it follows that $\angle OAB = \angle OBA$. Consider $\triangle OBA$: by [1.4], $\overline{OA} = \overline{OB}$. Similarly, we can show that

$$\overline{OA} = \overline{OB} = \overline{OC} = \overline{OD} = \overline{OE}$$

By [3.9], O is the center of a circle with radius $\overline{OA}$ which passes through points B, C, D, and E, and is constructed about the regular pentagon $ABCDE$. This completes the construction. $\square$

4.2. PROPOSITIONS FROM BOOK IV

Proposition 4.15. *INSCRIBING A REGULAR HEXAGON IN A CIRCLE.*

It is possible to inscribe a regular hexagon in a circle.

Proof. We wish to to inscribe regular hexagon $ABCDEF$ in $\odot O$.

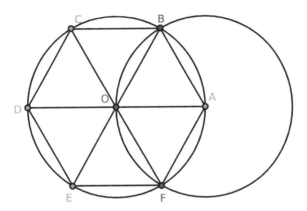

Figure 4.2.18: [4.15]

Take a point A on the circumference of $\odot O$ and construct $\overline{AO}$. With A as center and $\overline{AO}$ as radius, construct the circle $\odot A$, intersecting $\odot O$ at the points B and F.

Construct $\overline{OB}$ and $\overline{OF}$; extend $\overline{AO}$ to intersect $\odot A$ at D, extend $\overline{BO}$ to intersect $\odot A$ at E, and extend $\overline{FO}$ to intersect $\odot A$ at C. Also construct $\overline{AB}, \overline{BC}, \overline{CD}, \overline{DE}, \overline{EF}$, and $\overline{FA}$; we claim that hexagon $ABCDEF$ is the required hexagon.

Notice that $\overline{OA} = \overline{OB}$ since each are radii of $\odot O$. Similarly, $\overline{AB} = \overline{OA}$ since each are radii of $\odot A$. Hence, $\overline{OA} = \overline{OB} = \overline{AB}$, and so $\triangle OAB$ is equilateral.

Since the sum of the interior angles of a triangle is two right angles and equilateral triangles have equal angles, $\angle AOB = \angle OBA = \angle OAB$.

Mutatis mutandis, we can show that

$$\angle AOB = \angle AOF = \angle FOE = \angle EOD = \angle DOC = \angle BOC$$

By [Cor. 3.29.1], $\overline{AB} = \overline{BC} = \overline{CD} = \overline{DE} = \overline{EF} = \overline{FA}$ and so hexagon $ABCDEF$ is equilateral.

Also $\overline{OA} = \overline{OB} = \overline{OC} = \overline{OD} = \overline{OE} = \overline{OF}$ since each are radii of $\odot O$. By [1.8], each sub-triangle of hexagon $ABCDEF$ is congruent. It follows that

$$\angle ABC = \angle BCD = \angle CDE = \angle DEF = \angle EFA = \angle FAB$$

and so $ABCDEF$ is equiangular. This completes the construction. □

Remark. [4.13] and [4.14] are particular cases of the following theorem: "A regular polygon of any finite number of sides has one circle inscribed in it and another constructed about it, and both circles are concentric."

Corollary. *4.15.1. The length of the side of a regular hexagon inscribed in a circle is equal to the circle's radius.*

Corollary. *4.15.2. If three alternate angles of a hexagon are joined, they form an inscribed equilateral triangle.*

Exercises.

1. Prove that the area of a regular hexagon inscribed in a circle is equal to twice the area of an equilateral triangle inscribed in the circle. Also prove that the square of the side of the triangle equals three times the square of the side of the hexagon.

2. If the diameter of a circle is extended to C until the extended segment is equal to the radius, prove that the two tangents from C and their chord of contact form an equilateral triangle.

3. Prove that the area of a regular hexagon inscribed in a circle is half the area of an equilateral triangle and three-fourths of the area of a regular hexagon circumscribed to the circle.

4. Prove [Cor. 4.15.1].

5. Prove [Cor. 4.15.2].

4.2. PROPOSITIONS FROM BOOK IV

Proposition 4.16. *INSCRIBING A REGULAR FIFTEEN-SIDED POLYGON IN A GIVEN CIRCLE.*

It is possible to inscribe a regular fifteen-sided polygon in a given circle.

Proof. We wish to inscribe a regular fifteen-sided polygon in $\odot O$.

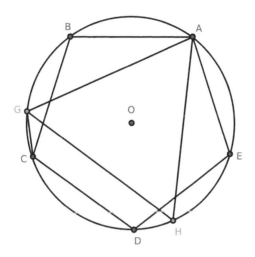

Figure 4.2.19: [4.16]

Inscribe a regular pentagon $ABCDE$ in the circle $\odot O$ [4.11] as well as an equilateral triangle $\triangle AGH$ [4.2]. Construct $\overline{CG}$. We claim that $\overline{CG}$ is a side of the required polygon.

Since $ABCDE$ is a regular pentagon, the arc ABC is $\frac{2}{5}$ of the circumference.

Since $\triangle AGH$ is an equilateral triangle, the arc ABG is $\frac{1}{3}$ of the circumference.

Hence arc $\overline{GC}$ is the difference between these two arcs and equal to $\frac{2}{5} - \frac{1}{3} = \frac{1}{15}$ of the circumference.

Therefore, if chords equal in length to $\overline{GC}$ are similarly constructed [4.1], we have a regular fifteen-sided polygon (i.e., a quindecagon) inscribed in $\odot O$. □

Remark. Until 1801, no regular polygon was constructible by segments and circles only except those described in Book IV of Euclid and those obtained by the continued bisection of the arcs of which their sides are chords. Then, Gauss proved that if $2n + 1$ is a prime number, regular polygons with $2n + 1$ sides are constructible by elementary geometric methods.

Exam questions for chapter 4.

1. What is the subject-matter of chapter 4?

2. When is one polygon said to be inscribed in another?

3. When is one polygon said to be circumscribed about another?

4. When is a circle said to be inscribed in a polygon?

5. When is a circle said to be circumscribed about a polygon?

6. What is a regular polygon?

7. What figures can be inscribed in, and circumscribed about, a circle by means of chapter 4?

8. What regular polygons has Gauss proved to be constructible by the line and circle?

9. What is meant by escribed circles?

10. How many circles can be constructed to touch three lines forming a triangle?

11. What is the centroid of a triangle?

12. What is the orthocenter?

13. What is the circumcenter?

14. What is the polar circle?

15. What is the "nine-points circle"?

16. How does a nine-points circle get its name?

17. Name the nine points that a nine-points circle passes through.

18. What three regular figures can be used in filling up the space round a point? (Ans. Equilateral triangles, squares, and hexagons.)

19. If the sides of a triangle are 13, 14, 15 units in length, what are the values of the radii of its inscribed and escribed circles?

20. What is the radius of the circumscribed circle? (See #19, above.)

21. What is the radius of its nine-points circle? (See #19, above.)

22. What is the distance between the centers of its inscribed and circumscribed circles? (See #19, above.)

23. If r is the radius of a circle, what is the area:

   (a) of its inscribed equilateral triangle?

   (b) of its inscribed square?

   (c) its inscribed pentagon?

4.2. PROPOSITIONS FROM BOOK IV

(d) its inscribed hexagon?

(e) its inscribed octagon?

(f) its inscribed decagon?

24. Find the side-lengths of the polygons in parts (a)-(f) in the previous problem.

Exercises for chapter 4.

1. If a circumscribed polygon is regular, prove that the corresponding inscribed polygon is also regular. Also prove the converse.

2. If a circumscribed triangle is isosceles, prove that the corresponding inscribed triangle is isosceles. Also prove the converse.

3. If the two isosceles triangles from #2 (above) have equal vertical angles, prove that they are both equilateral.

4. Divide an angle of an equilateral triangle into five equal parts.

5. Inscribe a circle in a sector of a given circle.

6. Inscribe a regular octagon in a given square.

7. If a segment of given length slides between two given lines, find the locus of the intersection of perpendiculars from its endpoints to the given lines.

8. If the perpendicular to any side of a triangle at its midpoint meets the internal and external bisectors of the opposite angle at the points D and E, prove that D, E are points on the circumscribed circle.

9. Through a given point P, construct a chord of a circle so that the intercept $\overline{EF}$ stands opposite a given angle at point X.

10. In a given circle, inscribe a triangle having two sides passing through two given points and the third parallel to a given line.

11. Given four points, no three of which are collinear, construct a circle which is equidistant from them.

12. In a given circle, inscribe a triangle whose three sides pass through three given points.

13. Construct a triangle, being given:

(a) the radius of the inscribed circle, the vertical angle, and the perpendicular from the vertical angle on the base.

(b) the base, the sum or difference of the other sides, and the radius of the inscribed circle, or of one of the escribed circles.

(c) the centers of the escribed circles.

14. If F is the midpoint of the base of a triangle, DE the diameter of the circumscribed circle which passes through F, and L the point where a parallel to the base through the vertex meets $\overline{DE}$, prove that $\overline{DL} \cdot \overline{FE}$ equals the square of half the sum of the two remaining sides and $\overline{DF} \cdot \overline{LE}$ equals the square of half the difference of the two remaining sides.

15. If from any point within a regular polygon of n sides perpendiculars fall on the sides, prove that their sum is equal to n times the radius of the inscribed circle.

16. The sum of the lengths of perpendiculars falling from the vertices of a regular polygon of n sides on any line is equal to n times the perpendicular from the center of the polygon on the same line.

17. If R denotes the radius of the circle circumscribed about a triangle $\triangle ABC$, r, r', r'', r''' are the radii of its inscribed and escribed circles; δ, δ', δ'' are the perpendiculars from its circumcenter on the sides; μ, μ', μ'' are the segments of these perpendiculars between the sides and circumference of the circumscribed circle, prove that we have the equalities:

$$\begin{aligned} r' + r'' + r''' &= 4R + r \quad (1) \\ \mu + \mu' + \mu'' &= 2R - r \quad (2) \\ \delta + \delta' + \delta'' &= R + r \quad (3) \end{aligned}$$

The relation (3) supposes that the circumcenter is inside the triangle.

18. Take a point D from the side BC of a triangle $\triangle ABC$ and suppose we construct a transversal EDF through it; suppose we also construct circles about the triangles $\triangle DBF$, $\triangle ECD$. Prove that the locus of their second point of intersection is a circle.

19. In every quadrilateral circumscribed about a circle, prove that the midpoints of its diagonals and the center of the circle are collinear.

20. Prove that the line joining the orthocenter of a triangle to any point P in the circumference of its circumscribed circle is bisected by the line of co-linearity of perpendiculars from P on the sides of the triangle.

21. Prove that the orthocenters of the four triangles formed by any four lines are collinear.

22. If a semicircle and its diameter are touched by any circle either internally or externally, prove that twice the area of the rectangle contained by the radius of the semicircle and the radius of the tangential circle equals the area of a rectangle contained by the segments of any secant to the semicircle through the point of intersection of the diameter and touching circle.

23. If ρ, ρ' are radii of two circles touching each other at the center of the inscribed circle of a triangle where each touches the circumscribed circle, prove that

4.2. PROPOSITIONS FROM BOOK IV

$$\frac{1}{\rho} + \frac{1}{\rho'} = \frac{2}{r}$$

and state and prove corresponding theorems for the escribed circles.

24. If from any point in the circumference of the circle, circumscribed about a regular polygon of n sides, segments are constructed to its vertices, prove that the sum of their squares is equal to $2n$ times the square of the radius.

25. In the above problem, if the segments are constructed from any point in the circumference of the inscribed circle, prove that the sum of their squares is equal to n times the sum of the squares of the radii of the inscribed and the circumscribed circles.

26. State the corresponding theorem for the sum of the squares of the lines constructed from any point in the circumference of any concentric circle.

27. If from any point in the circumference of any concentric circle perpendiculars fall to all the sides of any regular polygon, prove that the sum of their squares is constant.

28. See #27. For the inscribed circle, prove that the constant is equal to $3n/2$ times the square of the radius.

29. See #27. For the circumscribed circle, prove that the constant is equal to n times the square of the radius of the inscribed circle, together with $\frac{1}{2}n$ times the square of the radius of the circumscribed circle.

30. If from the midpoint of the segment joining any two of four concyclic points a perpendicular falls on the segment joining the remaining two, the six perpendiculars thus obtained are concurrent.

31. Given a regular polygon circumscribed about an arbitrary circle, prove that as the number of sides of a regular polygon increases, the perimeter of the polygon decreases.

32. The area of any regular polygon of more than four sides circumscribed about a circle is less than the square of the diameter.

33. If two sides of a triangle are given in position, and if their included angle is equal to an angle of an equilateral triangle, prove that the locus of the center of its nine-points circle is a straight line.

34. If s equals half of the perimeter of a triangle (i.e., the triangle's semi-perimeter), and if r', r'', r''' are the radii of its escribed circles, prove that

$$r' \cdot r'' + r'' \cdot r''' + r''' \cdot r' = s^2$$

35. Given the base of a triangle and the vertical angle, find the locus of the center of the circle passing through the centers of the escribed circles.

36. If $\overline{AB}$ is the diameter of a circle, $\overline{PQ}$ is any chord cutting $\overline{AB}$ at O, and if the segments $\overline{AP}$, $\overline{AQ}$ intersect the perpendicular to $\overline{AB}$ at O (at D and E respectively), prove that the points A, B, D, E are concyclic.

37. Inscribe in a given circle a triangle having its three sides parallel to three given lines.

38. If the sides $\overline{AB}$, $\overline{BC}$, etc., of a regular pentagon are bisected at the points A', B', C', D', E', and if the two pairs of alternate sides $\overline{BC}$, $\overline{AE}$ and $\overline{AB}$, $\overline{DE}$ meet at the points A'', E'', respectively, prove that

$$\triangle A''AE'' - \triangle A'AE' = \text{pentagon } A'B'C'D'E'$$

39. In a circle, prove that an equilateral inscribed polygon is regular; also prove that if the number of its sides is odd, then it is an equilateral circumscribed polygon.

40. Prove that an equiangular circumscribed polygon is regular; also prove that if the number of its sides are odd, then it is an equilateral inscribed polygon.

41. Prove that the sum of the perpendiculars constructed to the sides of an equiangular polygon from any point inside the figure is constant.

42. Express the lengths of the sides of a triangle in terms of the radii of its escribed circles.

Chapter 5

Theory of Proportions

Chapter 5, like Chapter 2, proves a number of propositions that are familiar in the form of algebraic equations. Like Book II, Book V appears in truncated form.

5.1 Definitions

0. Variables a, b, c, x, y, ... represent positive real numbers unless stated otherwise.

1. Let x and y be two positive integers where $x < y$. We say that x is a *factor* of y when there exists a positive integer $n \geq 1$ such that $nx = y$. We also say that y is a *multiple* of x.

2. Suppose that x and y are two positive integers such that $x \neq 0$. A *ratio* is the number $\frac{y}{x}$ which may also be written as $y : x$.

3. Numbers are said to have a *ratio* to one another when the lesser number can be multiplied so as to exceed the greater.

4. Numbers which have the same ratio are called *proportions*. When four numbers are proportions, it may be described as: "The first is to the second as the third is to the fourth." Or:

$$\frac{a}{b} = \frac{c}{d}$$

The above equality may also be written as $a : b = c : d$.

5. Inequalities of fractions:

$$\left(\frac{a}{b} > \frac{c}{d}\right) \Longleftrightarrow \left(\frac{ad}{bd} > \frac{bc}{bd}\right) \Longleftrightarrow (ad > bc)$$

The symbol > may be replaced with ≥, <, and ≤. Numbers which have the same ratio are called *proportional*.

6. Proportions consist of at minimum three terms.

Remark. This definition has the same fault as some of the others: it is not a definition but an inference. It occurs when the means in a proportion are equal, and so there are four terms. As an example, take the numbers 4, 6, 9. Here the ratio of $4:6$ is $2/3$, and the ratio of $6:9$ is $2/3$; therefore 4, 6, 9 are continued proportionals. But, in reality, there are four terms: the full proportion is $4:6=6:9$.

7. The duplicate ratio is a compound ratio of two equal ratios. Algebraically, the duplicate ratio of $x:y$ is $x^2:y^2$. (The duplicate ratio of $2:5$ is $4:25$.)

John Casey updates Euclid's definition of the duplicate ratio of two lines: *the ratio of the squares constructed on these segments.*

8. The triplicate ratio is a compound ratio of three equal ratios. Algebraically, the triplicate ratio of $x:y$ is $x^3:y^3$. (The triplicate ratio of $2:5$ is $8:125$.)

9. *Harmonic division* of a segment $\overline{AB}$ means identifying two points C and D such that $\overline{AB}$ is divided internally and externally in the same ratio $\frac{CA}{CB} = \frac{DA}{DB}$.

Figure 5.1.1: Here, the ratio is 2. Specifically, the distance $\overline{AC}$ is one unit, the distance $\overline{CB}$ is half a unit, the distance $\overline{AD}$ is three units, and the distance $\overline{BD}$ is 1.5 units.

Harmonic division of a line segment is reciprocal: if points C and D divide the segment $\overline{AB}$ harmonically, the points A and B also divide the line segment $\overline{CD}$ harmonically. In that case, the ratio is given by $\frac{BC}{BD} = \frac{AC}{AD}$ which equals one-third in the example above.[1]

---

[1] http://en.wikipedia.org/wiki/Harmonic_division

5.2 Propositions from Book V

Proposition 5.1. *If any number of numbers are each the same multiple of the same number of other numbers, then the sum is that multiple of the sum.*

Corollary. *5.1.1. [5.1] is equivalent to $kx + ky = k(x+y)$.*

Proposition 5.2. *If a first number is the same multiple of a second that a third is of a fourth, and a fifth also is the same multiple of the second that a sixth is of the fourth, then the sum of the first and fifth also is the same multiple of the second that the sum of the third and sixth is of the fourth.*

Corollary. *5.2.1. [5.2] is equivalent to the following: if $kv = x$, $kw = r$, $mv = y$, and $mw = u$, then $x + y = (k+m)v$ and $r + u = (k+m)w$.*

Proposition 5.3. *If a first number is the same multiple of a second that a third is of a fourth, and if equimultiples are taken of the first and third, then the numbers taken also are equimultiples respectively, the one of the second and the other of the fourth.*

Corollary. *5.3.1. [5.3] is equivalent to the following: Let $A = kB$ and $C = kD$. If $EF = mA$ and $GM = mC$, then $EF = mkB$ and $GH = mkD$.*

Proposition 5.4. *If a first number has to a second the same ratio as a third to a fourth, then any equimultiples whatever of the first and third also have the same ratio to any equimultiples whatever of the second and fourth respectively, taken in corresponding order.*

Corollary. *5.4.1. [5.4] is equivalent to: if $\frac{A}{B} = k = \frac{C}{D}$, then $A = kB$ and $C = kD$.*

Proposition 5.5. *If a number is the same multiple of a number that a subtracted part is of a subtracted part, then the remainder also is the same multiple of the remainder that the whole is of the whole.*

Corollary. *5.5.1. [5.5] is equivalent to: if $x + y = k(m+n)$ and $x = km$, then $y = kn$.*

Proposition 5.6. *If two numbers are equimultiples of two numbers, and any numbers subtracted from them are equimultiples of the same, then the remainders either equal the same or are equimultiples of them.*

Corollary. 5.6.1. *[5.6] is equivalent to: if $x + y = km$, $u + v = kn$, $x = lm$, $y = ln$, and all variables are positive, then $y = (k - l)m$ and $v = (k - l)n$ whenever $k > l$.*

Proposition 5.7. *Equal numbers have to the same the same ratio; and the same has to equal numbers the same ratio.*

Corollary. 5.7.1. *If any numbers are proportional, then they are also proportional inversely.*

Corollary. 5.7.2. *[5.7] is equivalent to: if $a = b$, then $a : c = b : c$, and $c : a = c : b$.*

Proposition 5.8. *Of unequal numbers, the greater has to the same a greater ratio than the less has; and the same has to the less a greater ratio than it has to the greater.*

Corollary. 5.8.1. *[5.8] is equivalent to: if $AB > C$ and $D > 0$, then $AB = C + k$ where $k > 0$, and $\frac{AB}{D} = \frac{C+k}{D} = \frac{C}{D} + \frac{k}{D} > \frac{C}{D}$. It follows that $\frac{D}{C} > \frac{D}{AB}$, since all quantities are positive.*

Proposition 5.9. *Numbers which have the same ratio to the same equal one another; and numbers to which the same has the same ratio are equal.*

Corollary. 5.9.1. *[5.9] is equivalent to: if $A = kC$ and $B = kC$, then $A = B$.*

Proposition 5.10. *Of numbers which have a ratio to the same, that which has a greater ratio is greater; and that to which the same has a greater ratio is less.*

Corollary. 5.10.1. *[5.10] is equivalent to: if $\frac{A}{C} > \frac{B}{C}$ and $C > 0$, then $A > B$.*

Proposition 5.11. *Ratios which are the same with the same ratio are also the same with one another.*

Corollary. 5.11.1. *[5.11] is equivalent to: if $\frac{A}{B} = \frac{C}{D}$ and $\frac{C}{D} = \frac{E}{F}$, then $\frac{A}{B} = \frac{E}{F}$. This is the transitive property for fractions.*

Proposition 5.12. *If any number of numbers are proportional, then one of the antecedents is to one of the consequents as the sum of the antecedents is to the sum of the consequents.*

5.2. PROPOSITIONS FROM BOOK V

Corollary. 5.12.1. *[5.12] is equivalent to: if $\frac{x}{a} = \frac{y}{b} = \frac{z}{c}$, then*

$$\frac{x}{a} = \frac{y}{b} = \frac{z}{c} = \frac{x+y+z}{a+b+c}$$

Proposition 5.13. *If a first number has to a second the same ratio as a third to a fourth, and the third has to the fourth a greater ratio than a fifth has to a sixth, then the first also has to the second a greater ratio than the fifth to the sixth.*

Corollary. 5.13.1. *[5.13] is equivalent to: if $A = kB$, $C = kD$, $C = lD$, $E = jF$ and $l > j$, then $k = l$ and so $k > j$.*

Proposition 5.14. *If a first number has to a second the same ratio as a third has to a fourth, and the first is greater than the third, then the second is also greater than the fourth; if equal, equal; and if less, less.*

Corollary. 5.14.1. *[5.14] is equivalent to: if $A = kB$, $C = kD$, $A > C$, and $k > 0$, then $kB = A > C = kD$ and so $B > D$. If $A < C$ and $B < D$, the result follows mutatis mutandis.*

Proposition 5.15. *Parts have the same ratio as their equimultiples.*

Corollary. 5.15.1. *[5.15] is equivalent to: if $AB = kC$, $DE = kF$, and $C = mF$, then $AB = kmF = mDE$.*

Proposition 5.16. *If four numbers are proportional, then they are also proportional alternately.*

Corollary. 5.16.1. *[5.16] is equivalent to: if $\frac{A}{B} = \frac{C}{D}$, then $\frac{A}{C} = \frac{B}{D}$.*

Proposition 5.17. *If numbers are proportional taken jointly, then they are also proportional taken separately.*

Corollary. 5.17.1. *[5.17] is equivalent to: if $x + y = ky$, $u + v = kv$, and $x = ly$, then $ly + y = ky$, or $l + 1 = k$. Thus $u + v = (l+1)v$, or $u = lv$.*

Proposition 5.18. *If numbers are proportional taken separately, then they are also proportional taken jointly.*

Corollary. *5.18.1. [5.18] is equivalent to: if $x = ky$, $u = kv$, and $x + y = ly$, then $(k+1)y = ly$ and so $k+1 = l$ and $u + v = kv = (k+1)v = lv$.*

Proposition 5.19. *If a whole is to a whole as a part subtracted is to a part subtracted, then the remainder is also to the remainder as the whole is to the whole.*

Corollary. *5.19.1. If numbers are proportional taken jointly, then they are also proportional in conversion.*

Corollary. *5.19.2. [5.19] is equivalent to: if $x + y = k(u+v)$ and $x = ku$, then $y = kv$.*

Remark. 5.19.3. *If $u : v$, the [5.16] [5.17], [5.18], and [5.19] show the following proportions to be equivalent[2]:*

1. $u : v = x : y$
2. $(u+v) : v = (x+y) : y$
3. $(u+v) : u = (x+y) : x$
4. $(u+v) : (x+y) = v : y$
5. $(u+v) : (x+y) = u : x$
6. $u : x = v : y$

(2)-(5) *also hold when $+$ is replaced by $-$.*

Proposition 5.20. *If there are three numbers, and others equal to them in multitude, which taken two and two are in the same ratio, and if the first is greater than the third, then the fourth is also greater than the sixth; if equal, equal, and; if less, less.*

Corollary. *5.20.1. [5.20] is equivalent to: let $A = kB$, $B = lC$, $D = kE$, $E = lF$, and $A > C$. We wish to show that $D > F$.*

Suppose that $A = c + m$, $m > 0$. Then $A = klC$, $D = klF$, and so $\frac{A}{C} = \frac{D}{F}$.

Now $\frac{A}{C} > 1$ since $A > C$. If $D = F$, $\frac{A}{C} = 1$; and if $D < F$, $\frac{A}{C} < 1$. Hence, $D > F$.

The remaining cases follow mutandis mutatis.

---

[2] David E. Joyce provides these results at:
aleph0.clarku.edu/~djoyce/java/elements/bookX/propX29.html

5.2. PROPOSITIONS FROM BOOK V

Proposition 5.21. *If there are three numbers, and others equal to them in multitude, which taken two and two together are in the same ratio, and the proportion of them is perturbed, then, if the first number is greater than the third, then the fourth is also greater than the sixth; if equal, equal; and if less, less.*

Corollary. *5.21.1. The result of [5.21] is the same as the result [5.20].*

Proposition 5.22. *If there are any number of numbers whatever, and others equal to them in multitude, which taken two and two together are in the same ratio, then they are also in the same ratio.*

Corollary. *5.22.1. [5.22] is equivalent to: if $A = kB$, $B = lC$, $D = kE$, and $E = lF$, then $A = klC$ and $D = klF$.*

Proposition 5.23. *If there are three numbers, and others equal to them in multitude, which taken two and two together are in the same ratio, and the proportion of them be perturbed, then they are also in the same ratio.*

Corollary. *5.23.1. The result of [5.23] is the same as the result of [5.22].*

Proposition 5.24. *If a first number has to a second the same ratio as a third has to a fourth, and also a fifth has to the second the same ratio as a sixth to the fourth, then the sum of the first and fifth has to the second the same ratio as the sum of the third and sixth has to the fourth.*

Corollary. *5.24.1. [5.24] is equivalent to: if $x = km$, $u = kn$, $y = lm$, and $v = ln$, then $x + y = km + lm = (k+l)m$ and $u + v = kn + ln = (k+l)n$.*

Proposition 5.25. *If four numbers are proportional, then the sum of the greatest and the least is greater than the sum of the remaining two.*

Corollary. *5.25.1. [5.25] is equivalent to: let $x + y = k(u+v)$, $k > 1$, and $x = ku$. Since $x + y = ku + kv$, $y = kv$; and since $k > 1$, $y > v$.*

Exam questions for chapter 5.

1. What is the subject-matter of this chapter?

2. When is one number said to be a multiple of another?

3. What is a measure?

4. What is the ratio of two commensurable numbers?

5. What is meant by the ratio of incommensurable numbers?

6. Give an illustration of the ratio of incommensurables.

7. What are the terms of a ratio called?

8. What is duplicate ratio?

9. Define triplicate ratio.

10. What is proportion? (Ans. Equality of ratios.)

11. When is a segment divided harmonically?

12. What are reciprocal ratios?

Chapter 5 exercises.

1. Prove that if four numbers are proportionals, the sum of the first and second is to their difference as the sum of the third and fourth is to their difference.

2. Prove that if four numbers are proportionals, their squares, cubes, etc., are proportionals. [See the final chapter for a solution.]

3. If two proportions have three terms of one respectively equal to three corresponding terms of the other, the remaining term of the first is equal to the remaining term of the second.

4. If three numbers are continual proportionals, prove that the first is to the third as the square of the difference between the first and second is to the square of the difference between the second and third.

5. If $\overline{AB}$ is cut harmonically at C and D and is bisected at O, prove that $\overline{OC}, \overline{OB}, \overline{OD}$ are continual proportionals.

6. Continuing from #5: if O' is the midpoint of $\overline{CD}$, prove that $(\overline{OO'})^2 = (\overline{OB})^2 + (\overline{OD})^2$.

7. Continuing from #5: prove that $\overline{AB} \cdot (\overline{AC} + \overline{AD}) = 2 \cdot \overline{AC} \cdot \overline{AD}$, or $\frac{1}{AC} + \frac{1}{AD} = \frac{2}{AB}$

8. Continuing from #5: prove that $\overline{CD} \cdot (\overline{AD} + \overline{BD}) = 2 \cdot \overline{AD} \cdot \overline{BD}$, or $\frac{1}{BD} + \frac{1}{AD} = \frac{2}{AC}$

9. Continuing from #5: prove that $\overline{AB} \cdot \overline{CD} = 2 \cdot \overline{AD} \cdot \overline{CB}$.

Chapter 6

Applications of Proportions

Recall that we write $\triangle GHI = \triangle JKL$ to indicate that the area of $\triangle GHI$ equals the area of $\triangle JKL$.

Similarly, if we wish to state that the area of $\triangle ABC$ divided by the area of $\triangle DEF$ equals the area of $\triangle GHI$ divided by the area of $\triangle JKL$, we may write either

$$\frac{\triangle ABC}{\triangle DEF} = \frac{\triangle GHI}{\triangle JKL}$$
or
$$\triangle ABC : \triangle DEF = \triangle GHI : \triangle JKL$$

6.1 Definitions

1. *Similar polygons* are polygons that have the same shape, or one that has the same shape as the mirror image of the other. More precisely, one polygon can be obtained from the other by uniformly scaling (enlarging or shrinking), possibly with additional translation, rotation and reflection. This means that either object can be re-scaled, re-positioned, and reflected so as to coincide precisely with the other object.

A modern perspective of similarity is to consider polygons similar if one appears congruent to the other when zoomed in or out at some level.[1]

Similar polygons agree in shape; if they also agree in size, then they are congruent. If polygons A and B are similar, we will denote this as $A \sim B$.

(a) When the shape of a figure is given, it is said to be given in species. Thus a triangle whose angles are given is given in species. Hence, similar figures are of the same species.

(b) When the size of a figure is given, it is said to be given in magnitude, such as a square whose side is of given length.

---
[1]Adapted from: https://en.wikipedia.org/wiki/Similarity_(geometry)

(c) When the place which a figure occupies is known, it is said to be given in position.

(d) Any two equilateral triangles are similar.

2. A segment is said to be cut at a point *in extreme and mean ratio* when the whole segment is to the greater segment as the greater segment is to the lesser segment. This ratio is also referred to as the *golden ratio*.

The figure below illustrates the geometric relationship.

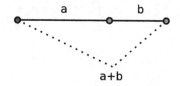

Figure 6.1.1: The golden ratio

Expressed algebraically, for quantities a and b with $a > b > 0$,

$$\frac{a+b}{a} = \frac{a}{b} \stackrel{\text{def}}{=} \varphi$$

where the Greek letter phi (φ or ϕ) represents the golden ratio. Its value is:

$$\varphi = \frac{1 + \sqrt{5}}{2} = 1.6180339887\ldots$$

The golden ratio also is called the *golden mean* or *golden section* (Latin: *sectio aurea*). Other names include *medial section*, *divine proportion*, *divine section* (Latin: *sectio divina*), *golden proportion*, *golden cut*, and *golden number*.[2]

3. If three quantities of the same kind are in continued proportion, the middle term is called a *mean proportional* between the other two. Numbers in continued proportion are also said to be in *geometrical progression*.

4. If four quantities of the same kind are in continued proportion, the two middle terms are called two *mean proportionals* between the other two.

5. The *altitude* of any figure is the length of the perpendicular from its highest point to its base.

6. Two corresponding angles of two figures have the sides about them *reciprocally proportional* when a side of the first is to a side of the second as the remaining side of the second is to the remaining side of the first.

---

[2]Much of this section comes from https://en.wikipedia.org/wiki/Golden_ratio.

6.1. DEFINITIONS

7. *Similar figures* are said to be similarly constructed upon given segments when these segments are corresponding sides of the figures.

8. *Corresponding points* in the planes of two similar figures are such that segments constructed from them to the vertices of the two figures are proportional to the corresponding sides of the two figures. See Fig. 6.1.2.

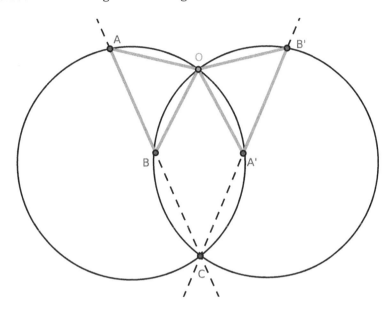

Figure 6.1.2: [Def 6.9] See [6.20, #2]

9. The point O in Fig. 6.1.2 is called the *center of similitude of the figures*. It is also called their double point.

10. Two polygons are said to be homothetic if they are similar and their corresponding sides are parallel. If two polygons are homothetic, then the lines joining their corresponding vertices meet at a point.[3]

11. The center of mean position of any number of points A, B, C, D, etc., is a point which may be found as follows: bisect the segment joining any two points A, B at G. Join G to a third point C; divide $\overline{GC}$ at H so that $\overline{GH} = \frac{1}{3} \cdot \overline{GC}$. Join H to a fourth point D and divide $\overline{HD}$ at K, so that $\overline{HK} = \frac{1}{4} \cdot \overline{HD}$, and so on. The last point found will be the center of mean position of the given points.

---

[3]http://americanhistory.si.edu/collections/search/object/nmah_694635

6.2 Propositions from Book VI

Proposition 6.1. *PROPORTIONAL TRIANGLES AND PARALLELOGRAMS.*

Triangles and parallelograms which have the same altitude have areas which are proportional to their bases.

Proof. Construct $\triangle ACB$ and $\triangle ACD$ such that the triangles have the same altitude; also construct $\square EACB$ and $\square AFDC$ each with the same height as the previously constructed triangles. We claim that

$$\overline{BC}:\overline{CD} = \triangle ACB : \triangle ACD = \square EACB : \square AFDC$$

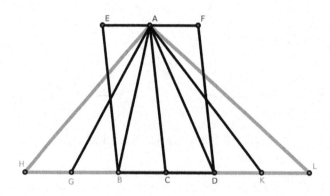

Figure 6.2.1: [6.1]

Extend $\overline{BD}$ in both directions to the H and L. Construct any finite number of segments toward H, each equal in the length to $\overline{BC}$: in this proof, we construct $\overline{BG}$ and $\overline{GH}$. Similarly, construct an equal number of segments which are equal in the length to $\overline{CD}$ toward L. Also construct $\overline{AG}$, $\overline{AH}$, $\overline{AK}$, and $\overline{AL}$.

Since $\overline{BC} = \overline{BG} = \overline{GH}$, by [1.38] $\triangle ACB = \triangle ABG = \triangle AGH$. Similarly, if $\overline{CH} = k \cdot \overline{BC}$ (where k is a positive integer such that $k > 1$), $\triangle ACH = k \cdot \triangle ACB$; and if $\overline{CL} = m \cdot \overline{CD}$ (where m is a positive integer such that $m > 1$), $\triangle ACL = m \cdot \triangle ACD$.

Finally, $\overline{CH} = n \cdot \overline{CL}$ (where n is a positive real number such that $n > 0$); again, [1.38] implies that $\triangle ACH = n \cdot \triangle ACL$. That is,

$$\frac{CH}{CL} = n = \frac{\triangle ACH}{\triangle ACL}$$

Hence,

6.2. PROPOSITIONS FROM BOOK VI

$$
\begin{aligned}
\overline{CH} : \overline{CL} &= \triangle ACH : \triangle ACL & \Rightarrow \\
k \cdot \overline{BC} : m \cdot \overline{CD} &= (k \cdot \triangle ACB) : (m \cdot \triangle ACD) & \Rightarrow \\
\overline{BC} : \overline{CD} &= \triangle ACB : \triangle ACD &
\end{aligned}
$$

By [1.41], $\square EACB = 2 \cdot \triangle ACB$ and $\square ACDF = 2 \cdot \triangle ACD$. Therefore

$$
\begin{aligned}
2 \cdot \overline{BC} : 2 \cdot \overline{CD} &= \square EACB : \square AFCD & \Rightarrow \\
\overline{BC} : \overline{CD} &= \square EACB : \square AFCD &
\end{aligned}
$$

Clearly, $\triangle ACB : \triangle ACD = \square EACB : \square AFCD$, which completes the proof. □

Proposition 6.2. *PROPORTIONALITY OF SIDES OF TRIANGLES.*

A segment within a triangle divides the sides of a triangle proportionally if and only if the segment is constructed parallel to one of the sides of a triangle.

Proof. Construct $\triangle ABC$ and $\overline{DE}$ where D is on $\overline{AB}$ and E is on $\overline{AC}$. We claim that $\overline{DE} \parallel \overline{BC}$ if and only if $\overline{AD} : \overline{DB} = \overline{AE} : \overline{EC}$.

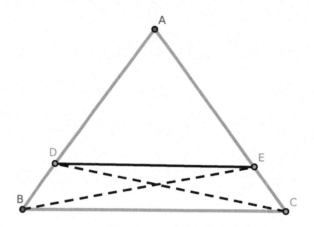

Figure 6.2.2: [6.2]

First, suppose that $\overline{DE} \parallel \overline{BC}$. We claim that $\overline{AD} : \overline{DB} = \overline{AE} : \overline{EC}$.

Construct $\overline{BE}$ and $\overline{CD}$, and consider $\triangle BDE$ and $\triangle CED$: each shares base $\overline{DE}$ and stands between the parallels $\overline{BC}$ and $\overline{DE}$. By [1.37], $\triangle BDE = \triangle CDE$. By [5.7], $\triangle ADE : \triangle BDE = \triangle ADE : \triangle CDE$.

By [6.1], $\triangle ADE : \triangle BDE = \overline{AD} : \overline{DB}$ and $\triangle ADE : \triangle CDE = \overline{AE} : \overline{EC}$. Since $\triangle BDE = \triangle CDE$, it follows that $\overline{AD} : \overline{DB} = \overline{AE} : \overline{EC}$.

Now suppose that $\overline{AD} : \overline{DB} = \overline{AE} : \overline{EC}$. We claim $\overline{DE} \parallel \overline{BC}$.

By [6.1], $\overline{AD} : \overline{DB} = \triangle ADE : \triangle BDE$ and $\overline{AE} : \overline{EC} = \triangle ADE : \triangle CDE$. Since $\overline{AD} : \overline{DB} = \overline{AE} : \overline{EC}$ by hypothesis, $\triangle ADE : \triangle BDE = \triangle ADE : \triangle CDE$.

By [5.9], $\triangle BDE = \triangle CDE$. These triangles also stand on the same base $\overline{DE}$ as well as on the same side of $\overline{DE}$. By [1.39], they stand between the same parallels, and so $\overline{DE} \parallel \overline{BC}$. □

Exercise.

1. If two segments are cut by three or more parallels, the intercepts on one are proportional to the corresponding intercepts on the other.

6.2. PROPOSITIONS FROM BOOK VI

Proposition 6.3. *ANGLES AND PROPORTIONALITY OF TRIANGLES.*

A line bisects an angle of a triangle if and only if the line divides the side opposite the angle into segments proportional to the adjacent sides.

Proof. Construct $\triangle ABC$ and $\overleftrightarrow{AD}$ where D is a point on $\overline{AC}$. We claim that $\overline{AD}$ bisects $\angle BAC$ if and only if $\overline{BD} : \overline{DC} = \overline{BA} : \overline{AC}$.

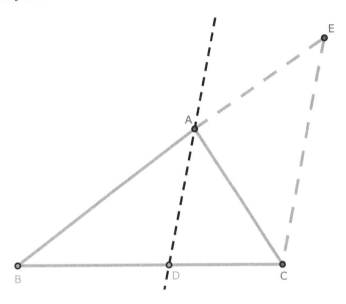

Figure 6.2.3: [6.3]

Suppose that $\overline{AD}$ bisects $\angle BAC$ of a triangle $\triangle ABC$. We claim that $\overline{BD} : \overline{DC} = \overline{BA} : \overline{AC}$.

Construct $\overline{CE}$ such that $\overline{CE} \parallel \overline{AD}$. Extend $\overline{BA}$ to intersect $\overline{CE}$ at E. Because $\overline{BE}$ intersects the parallels $\overline{AD}$ and $\overline{EC}$, $\angle BAD = \angle AEC$ [1.29].

Because $\overline{AC}$ intersects the parallels $\overline{AD}$ and $\overline{EC}$, $\angle ACE = \angle DAC$. By hypothesis, $\angle DAC = \angle BAD$. Therefore, $\angle ACE = \angle DAC = \angle BAD = \angle AEC$. Consider $\triangle ACE$: since $\angle ACE = \angle AEC$, by [1.6] $\overline{AE} = \overline{AC}$.

Again, because $\overline{AD} \parallel \overline{EC}$, where EC is one of the sides of the triangle $\triangle BEC$, by [6.2] $\overline{BD} : \overline{DC} = \overline{BA} : \overline{AE}$. Since $\overline{AE} = \overline{AC}$ by the above, $\overline{BD} : \overline{DC} = \overline{BA} : \overline{AC}$.

Now suppose that $\overline{BD} : \overline{DC} = \overline{BA} : \overline{AC}$. We claim that $\angle BAC$ is bisected by $\overline{AD}$.

Let the same construction be made as above. Because $\overline{AD} \parallel \overline{EC}$, by [6.2] $\overline{BA} : \overline{AE} = \overline{BD} : \overline{DC}$.

But $\overline{BD} : \overline{DC} = \overline{BA} : \overline{AC}$ by hypothesis. By [5.11], it follows that $\overline{BA} : \overline{AE} = \overline{BA} : \overline{AC}$, and so $\overline{AE} = \overline{AC}$ by [5.9].

Consider $\triangle ACE$: by [Cor. 1.6.1], $\angle AEC = \angle ACE$. By [1.29], $\angle AEC = \angle BAD$ and $\angle ACE = \angle DAC$. Hence $\angle BAD = \angle DAC$.

Since $\angle BAC = \angle BAD + \angle DAC$, it follows that $\angle BAC$ is bisected by $\overline{AD}$. This completes the proof. □

Exercises.

1. If the segment $\overline{AD}$ bisects the external vertical angle $\angle CAE$, prove that $\overline{BA} : \overline{AC} = \overline{BD} : \overline{DC}$, and conversely.

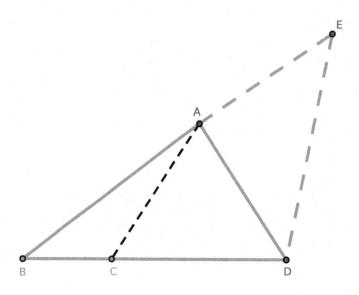

Figure 6.2.4: [6.3], #1

Hint: construct $\overline{AE} = \overline{AC}$. Also construct $\overline{ED}$. Then the triangles $\triangle ACD$ and $\triangle AED$ are evidently congruent; therefore the angle $\angle EDB$ is bisected, and hence $\overline{BA} : \overline{AE} = \overline{BD} : \overline{DE}$ and $\overline{BA} : \overline{AC} = \overline{BD} : \overline{DC}$ by [6.3].

2. Prove #1 without using [6.3], and then prove [6.3] using #1.

3. Prove that the internal and the external bisectors of the vertical angle of a triangle divide the base harmonically.

4. Prove that any segment intersecting the legs of any angle is cut harmonically by the internal and external bisectors of the angle.

5. Prove that any segment intersecting the legs of a right angle is cut harmonically by any two lines through its vertex which make equal angles with either of its sides.

6. If the base of a triangle is given in number and position and if the ratio of the sides is also given, prove that the locus of the vertex is a circle which divides the base harmonically in the ratio of the sides.

6.2. PROPOSITIONS FROM BOOK VI

7. If a, b, c denote the sides of a triangle $\triangle ABC$, and D, D' are the points where the internal and external bisectors of A meet BC, then prove that

$$DD' = \frac{2abc}{b^2 - c^2}$$

8. In the same case as #7, if E, E', F, F' are points similarly determined on the sides CA, AB, respectively, prove that

$$\frac{1}{DD'} + \frac{1}{EE'} + \frac{1}{FF'} = 0$$
$$\frac{a^2}{DD'} + \frac{b^2}{EE'} + \frac{c^2}{FF'} = 0$$

Proposition 6.4. *EQUIANGULAR TRIANGLES I.*

In equiangular triangles, the sides about the equal angles are proportional where the corresponding sides stand opposite the equal angles.

Proof. Let $\triangle ABC$ and $\triangle DCE$ be equiangular triangles where $\angle ABC = \angle DCE$, $\angle BAC = \angle CDE$, and $\angle ACB = \angle CED$. We claim that $\overline{BA} : \overline{AC} = \overline{CD} : \overline{DE}$. (The proof of the remaining cases will be analogous.)

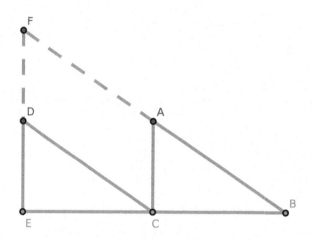

Figure 6.2.5: [6.4]

Place $\overline{BC}$ such that $\overline{BE} = \overline{BC} \oplus \overline{CE}$. Since $\angle ABC + \angle ACB$ is less than two right angles [1.17] and $\angle ACB = \angle DEC$, it follows that $\angle ABC + \angle DEC$ is less than two right angles. Hence when $\overline{BA}$ and $\overline{DE}$ are extended, they will intersect at F by [Cor. 1.29.1].

By [1.28], since $\angle DCE = \angle ABC$, $\overline{DC} \parallel \overline{FB}$; and since $\angle DEC = \angle ACB$, $\overline{AC} \parallel \overline{FE}$. Therefore $\square FACD$ is a parallelogram; by [1.34], $\overline{FA} = \overline{DC}$ and $\overline{AC} = \overline{FD}$.

Consider $\triangle FBE$: since $\overline{AC} \parallel \overline{FE}$, by [6.2] $\overline{BA} : \overline{AF} = \overline{BC} : \overline{CE}$. But $\overline{AF} = \overline{CD}$, and so $\overline{BA} : \overline{CD} = \overline{BC} : \overline{CE}$. By [5.16], $\overline{BA} : \overline{BC} = \overline{CD} : \overline{CE}$.

Similarly, since $\overline{CD} \parallel \overline{BF}$, by [6.2] $\overline{BC} : \overline{CE} = \overline{FD} : \overline{DE}$. But $\overline{FD} = \overline{AC}$, and so $\overline{BC} : \overline{CE} = \overline{AC} : \overline{DE}$; or $\overline{BC} : \overline{AC} = \overline{CE} : \overline{DE}$.

Since $\overline{BA} : \overline{BC} = \overline{CD} : \overline{CE}$ and $\overline{BC} : \overline{AC} = \overline{CE} : \overline{DE}$, by [5.22] $\overline{BA} : \overline{AC} = \overline{CD} : \overline{DE}$, which proves our claim. □

Corollary. *6.4.1. By [Def. 6.1], the triangles in [6.4] have been proved to be similar. Therefore, equiangular triangles are similar.*

6.2. PROPOSITIONS FROM BOOK VI

Exercises.

1. If two circles intercept equal chords $\overline{AB}$, $\overline{A'B'}$ on any secant, prove that the tangents $\overleftrightarrow{AT}$, $\overleftrightarrow{A'T}$ to the circles at the points of intersection are to one another as the radii of the circles.

2. If two circles intercept on any secant chords that have a given ratio, prove that the tangents to the circles at the points of intersection have a given ratio, namely, the ratio compounded of the direct ratio of the radii and the inverse ratio of the chords.

3. Being given a circle and a line, prove that a point may be found such that the rectangle of the perpendiculars falling on the line from the points of intersection of the circle with any chord through the point shall be given.

4. If $\overline{AB}$ is the diameter of a semicircle ADB and $\overline{CD} \perp \overline{AB}$, construct through A a chord $\overline{AF}$ of the semicircle meeting $\overline{CD}$ at E such that the ratio $\overline{CE} : \overline{EF}$ may be given.

Proposition 6.5. *EQUIANGULAR TRIANGLES II.*

If two triangles have proportional sides, then the triangles are equiangular with the equal angles opposite the corresponding sides.

Proof. Construct $\triangle ABC$ and $\triangle DEF$ such that $\overline{BA} : \overline{AC} = \overline{ED} : \overline{DF}$ and $\overline{AC} : \overline{CB} = \overline{DF} : \overline{FE}$. We claim that:

(1) $\triangle ABC$ and $\triangle DEF$ are equiangular;

(2) equal angles stand opposite corresponding sides.

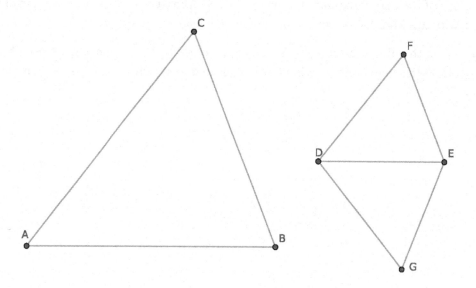

Figure 6.2.6: [6.5]

At D and E, construct the angles $\angle EDG = \angle BAC$ and $\angle DEG = \angle ABC$. By [1.32], $\triangle ABC$ and $\triangle DEG$ are equiangular; by [6.4], $\overline{BA} : \overline{AC} = \overline{ED} : \overline{DG}$.

Since $\overline{BA} : \overline{AC} = \overline{ED} : \overline{DF}$ by hypothesis, $\overline{DG} = \overline{DF}$. Similarly, $\overline{EG} = \overline{EF}$.

Consider $\triangle EDF$ and $\triangle EDG$: $\overline{DG} = \overline{DF}$, each shares side $\overline{ED}$, and $\overline{EG} = \overline{EF}$. By [1.8], $\triangle EDF \cong \triangle EDG$, and so $\triangle EDF$ and $\triangle EDG$ are equiangular. But $\triangle EDG$ is equiangular to $\triangle ABC$ by construction. Therefore, $\triangle EDF$ is equiangular to $\triangle ABC$, proving claim 1.

Since $\angle BAC$ stands between $\overline{BA}$ and $\overline{AC}$, and $\angle EDF$ stands between $\overline{ED}$ and $\overline{DF}$, we have also proven claim 2. This completes the proof. $\square$

Corollary. *6.5.1. Two triangles are equiangular if and only if the sides about the equal angles are proportional where the corresponding sides stand opposite the equal angles.*

6.2. PROPOSITIONS FROM BOOK VI

Remark. In [Def. 6.1], two conditions are laid down as necessary for the similitude of polygons:

(a) The equality of angles;

(b) The proportionality of sides.

Now by [6.4] and [6.5], we see that if two triangles possess either condition, they also possess the other. Triangles are unique in this respect. In all other polygons, one of these conditions may exist without the other. Thus two quadrilaterals may have their sides proportional without having equal angles, or vice verse.

Proposition 6.6. *EQUIANGULAR TRIANGLES III.*

If two triangles contain an equal angle enclosed by proportional sides, then the triangles are equiangular and have those angles equal which stand opposite to their corresponding sides.

Proof. Construct △ABC and △DEF such that:

(1) ∠BAC = ∠EDF

(2) $\overline{BA} : \overline{AC} = \overline{ED} : \overline{DF}$

We claim that △ABC and △DEF are equiangular and have those angles equal which stand opposite to their corresponding sides.

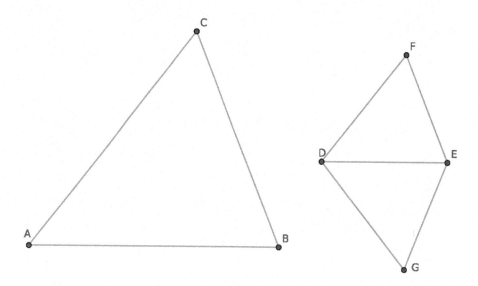

Figure 6.2.7: [6.6]

Recreate the construction from [6.5]: by [6.4], $\overline{BA} : \overline{AC} = \overline{ED} : \overline{DG}$. By hypothesis, $\overline{BA} : \overline{AC} = \overline{ED} : \overline{DF}$, and so $\overline{DG} = \overline{DF}$.

Because ∠EDG = ∠BAC by construction and ∠BAC = ∠EDF by hypothesis, ∠EDG = ∠EDF.

Consider △EDG and △EDF: $\overline{DG} = \overline{DF}$, each shares side $\overline{DE}$, and ∠EDG = ∠EDF. By [1.4], △EDF ≅ △EDG, and so △EDF and △EDG are equiangular.

But △EDG is equiangular to △BAC by construction, and so △EDF is equiangular to △BAC.

Finally, △ABC and △DEF have equal angles which stand opposite to their corresponding sides, which proves our claim. □

6.2. PROPOSITIONS FROM BOOK VI

Remark. As in the case of [6.4], an immediate proof of [6.6] can also be obtained from [6.2].

Proposition 6.7. *SIMILAR TRIANGLES I.*

If two triangles each contain an equal angle, if the sides about two remaining angles are proportional, and if the remaining angles are either both acute or not acute, then the triangles are similar.

Proof. Construct $\triangle ABC$ and $\triangle DEF$ where $\angle BAC = \angle EDF$, $\overline{AB} : \overline{BC} = \overline{DE} : \overline{EF}$, and $\angle EFD$ and $\angle BCA$ are either both acute or not acute. We claim that $\triangle ABC \sim \triangle DEF$.

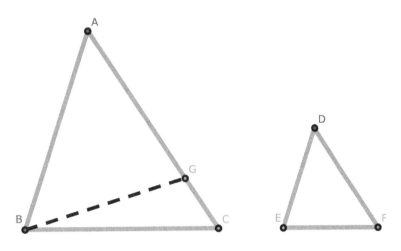

Figure 6.2.8: [6.7]

If $\angle ABC \neq \angle DEF$, then one angle must be greater than the other. Wlog, suppose $\angle ABC > \angle DEF$ such that $\angle ABG = \angle DEF$, from which it follows that $\angle EFD = \angle BGA$.

Consider $\triangle ABG$ and $\triangle DEF$: $\angle BAG = \angle EDF$ and $\angle ABG = \angle DEF$. By [1.32], $\triangle ABG$ and $\triangle DEF$ are equiangular. By [6.4], $\overline{AB} : \overline{BG} = \overline{DE} : \overline{EF}$.

Since $\overline{AB} : \overline{BC} = \overline{DE} : \overline{EF}$ by hypothesis, $\overline{BG} = \overline{BC}$. Consider $\triangle CBG$: by the above, the triangle is isosceles. By [1.5], $\angle BGC = \angle BCG$. By [Cor. 1.17.1], every triangle has at least two acute angles, and so $\angle BGC$ and $\angle BCG$ are both acute. Since $\angle BGC$ is acute, $\angle BGA$ is obtuse.

By hypothesis, $\angle EFD = \angle BGA$, and so $\angle EFD$ is obtuse; since $\angle BCG = \angle BCA$, $\angle BCA$ is acute. However, $\angle EFD$ and $\angle BCA$ are either both acute or both obtuse by hypothesis, a contradiction. Hence $\angle ABC = \angle DEF$. Since $\angle BAC = \angle EDF$ by hypothesis, by [1.32] $\triangle ABC$ and $\triangle EFD$ are equiangular.

Since $\overline{AB} : \overline{BC} = \overline{DE} : \overline{EF}$ by hypothesis, $\triangle ABC \sim \triangle DEF$ [6.4]. This proves our claim. □

Corollary. *6.7.1. If $\triangle ABC$ and $\triangle DEF$ each have two sides proportional to two sides in the other triangle, then $\overline{AB} : \overline{BC} = \overline{DE} : \overline{EF}$, the angles at points A and D opposite one pair of corresponding sides are equal, and the angles at points C and F opposite the other are either equal or supplemental. This proposition is nearly identical with [6.7].*

Corollary. *6.7.2. If either of the angles at points C and F are right, the other angle must be right.*

Exercises.

1. Prove [Cor. 6.7.1].

2. Prove [Cor. 6.7.2].

3. Prove the Transitivity of Similar Triangles, i.e., if $\triangle ABC \sim \triangle DEF$ and $\triangle DEF \sim \triangle GHI$, then $\triangle ABC \sim \triangle GHI$. [See the final chapter for the solution.]

6.2. PROPOSITIONS FROM BOOK VI

Proposition 6.8. *SIMILARITY OF RIGHT TRIANGLES.*

The triangles formed by dividing a right triangle by the perpendicular from the right angle to the hypotenuse are similar to the original triangle and to each other.

Proof. Construct right-triangle $\triangle ACB$ where $\angle BCA$ is its right angle. Construct $\overline{CD}$ such that D is on side AB and $\overline{AB} \perp \overline{DC}$. We claim that $\triangle ABC \sim \triangle ACD$, $\triangle ABC \sim \triangle BCD$, and $\triangle ACD \sim \triangle BCD$.

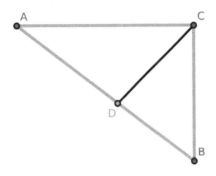

Figure 6.2.9: [6.8]

Consider $\triangle ABC$ and $\triangle ACD$: they share $\angle BAC$, and they each contain a right angle. By [1.32], $\triangle ABC$ and $\triangle ABD$ are equiangular. By [Cor. 6.4.1], $\triangle ABC \sim \triangle ABD$.

Likewise, $\triangle ABC \sim \triangle BCD$. By [6.7, #3], $\triangle ACD \sim \triangle BCD$, which completes the proof. $\square$

Corollary. 6.8.1. *The perpendicular $\overline{DC}$ is a mean proportional between the segments $\overline{AD}$ and $\overline{DB}$ of the hypotenuse. (Since $\triangle ADC$ and $\triangle CDB$ are equiangular, $\overline{AD} : \overline{DC} = \overline{DC} : \overline{DB}$. Thus $\overline{DC}$ is a mean proportional between $\overline{AD}$ and $\overline{DB}$ [Def. 6.3].)*

Corollary. 6.8.2. $\overline{BC}$ *is a mean proportional between* $\overline{AB}$ *and* $\overline{BD}$; *also,* $\overline{AC}$ *is a mean proportional between* $\overline{AB}$ *and* $\overline{AD}$.

Corollary. 6.8.3. *The segments $\overline{AD}$ and $\overline{DB}$ are in the duplicate of $\overline{AC} : \overline{CB}$; in other words, $\overline{AD} : \overline{DB} = (\overline{AC})^2 : (\overline{CB})^2$.*

Corollary. 6.8.4. $\overline{BA} : \overline{AD}$ *are in the duplicate ratios of* $\overline{BA} : \overline{AC}$, *and* $\overline{AB} : \overline{BD}$ *are in the duplicate ratio of* $\overline{AB} : \overline{BC}$. *Or,* $\overline{AB} : \overline{AD} = (\overline{AB})^2 : (\overline{AC})^2$ *and* $\overline{AB} : \overline{BD} = (\overline{AB})^2 : (\overline{BC})^2$.

Exercises.

1. Prove [Cor. 6.8.1].

2. Prove [Cor. 6.8.2].

3. Prove [Cor. 6.8.3].

4. Prove [Cor. 6.8.4].

Proposition 6.9. *CUTTING OFF SUB-SEGMENTS.*

From a given segment, we may cut off any required sub-segment.

Proof. Construct $\overline{AB}$; we wish to cut off any required sub-segment from $\overline{AB}$.

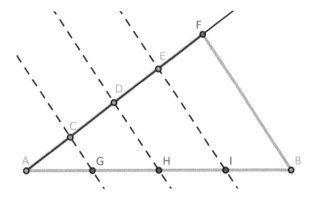

Figure 6.2.10: [6.9]

Suppose we wish to cut $\frac{1}{4}$th of $\overline{AB}$. Construct $\overrightarrow{AF}$ at any acute angle to $\overline{AB}$. On $\overrightarrow{AF}$, choose point C and cut off segments $\overline{CD}$, $\overline{DE}$, and $\overline{EF}$ where $\overline{AC} = \overline{CD} = \overline{DE} = \overline{EF}$ [1.3]. Construct $\overline{CG}$, $\overline{DH}$, $\overline{EI}$, and $\overline{FB}$ such that each is parallel to $\overline{CG}$. We claim that $\overline{AG} = \frac{1}{4} \cdot \overline{AB}$.

Since $\overline{CG} \parallel \overline{BF}$ where $\overline{BF}$ is the side of $\triangle ABF$, by [6.2]:

$$\begin{aligned}
\overline{CF} : \overline{AC} &= \overline{GB} : \overline{AG} \\
(\overline{AC} \oplus \overline{CF}) : \overline{AC} &= (\overline{AG} \oplus \overline{GB}) : \overline{AG} \\
\overline{AF} : \overline{AC} &= \overline{AB} : \overline{AG} \\
\overline{AC} : \overline{AF} &= \overline{AG} : \overline{AB}
\end{aligned}$$

But $\overline{AC} = \frac{1}{4} \cdot \overline{AF}$ by construction, and so $\overline{AG} = \frac{1}{4} \cdot \overline{AB}$. Since our choice of $\frac{1}{4}$ was arbitrary, any other required sub-segment may similarly be cut off. □

Remark. [1.10] is a particular case of this proposition.

Exercises.

1. Prove [6.9] using a proof by induction.

Proposition 6.10. *SIMILARLY DIVIDED SEGMENTS.*

We wish to divide a segment similarly to a given divided segment.

Proof. Construct $\overline{AC}$ divided at points D and E as well as $\overline{AB}$ where $\angle BAC$ is an acute angle. We claim that $\overline{AB}$ can be divided similarly to $\overline{AC}$.

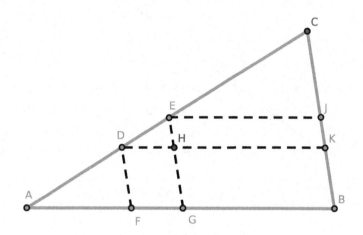

Figure 6.2.11: [6.10]

Construct $\overline{CB}$, $\overline{DF}$, and $\overline{EG}$ such that $\overline{DF} \parallel \overline{CB}$ and $\overline{EG} \parallel \overline{CB}$. Also construct $\overline{DK}$ where $\overline{DK}$ intersects $\overline{EG}$ at H and where $\overline{DK} \parallel \overline{AB}$.

It follows that $\square DHGF$ and $\square HKBG$ are parallelograms; by [1.34] $\overline{DH} = \overline{FG}$ and $\overline{HK} = \overline{GB}$.

Consider $\triangle DCK$: since $\overline{EH} \parallel \overline{CK}$, by [6.2] $\overline{DE} : \overline{EC} = \overline{DH} : \overline{HK}$.

But $\overline{DH} = \overline{FG}$ and $\overline{HK} = \overline{GB}$; by [5.7], we find that $\overline{DE} : \overline{EC} = \overline{FG} : \overline{GB}$.

Consider $\triangle AEG$: since $\overline{DF} \parallel \overline{EG}$, we find that $\overline{AD} : \overline{DE} = \overline{AF} : \overline{FG}$.

From the last two proportions we obtain $\overline{AD} : \overline{EC} = \overline{AF} : \overline{GB}$. This divides $\overline{AB}$ similarly to $\overline{AC}$. □

Corollary. *6.10.1. We may divide a given undivided segment $\overline{AB}$ similarly to a given divided segment $\overline{DE}$ by constructing $\overline{AC}$ at an acute angle to $\overline{AB}$ where $\overline{AC}$ is divided into segments similar to $\overline{DE}$.*

6.2. PROPOSITIONS FROM BOOK VI

Exercises.

1. We wish to divide a given segment $\overline{AB}$ internally or externally in the ratio of two given segments $\overline{FG}$ and $\overline{HJ}$.

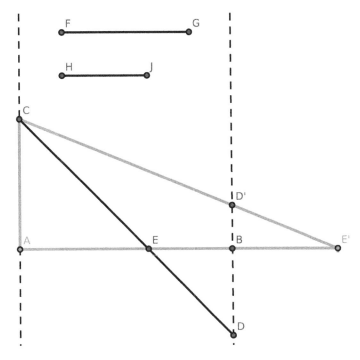

Figure 6.2.12: [6.11]

Through A and B construct any two parallels $\overleftrightarrow{AC}$ and $\overleftrightarrow{BD}$. Construct segments $\overline{AC} = \overline{FG}$ and $\overline{BD} = \overline{HJ}$. Also construct $\overline{CD}$: we claim that $\overline{CD}$ divides $\overline{AB}$ internally at E in the ratio of $\overline{FG} : \overline{HJ}$.

2. In #1, if $\overline{BD'}$ is constructed parallel to $\overline{AC}$, then $\overline{CD}$ will cut $\overline{AB}$ externally at E in the ratio of $\overline{FG} : \overline{HJ}$.

Corollary. *6.10.2. The two points in the above Figure, E and E', divide $\overline{AB}$ harmonically.*

This problem is manifestly equivalent to the following: given the sum or difference of two segments and their ratio, we wish to construct the segments.

Exercises.

3. In the above Figure, prove that any line $\overline{AE'}$ through the midpoint B of the base $\overline{DD'}$ of $\triangle DCD'$ is cut harmonically by the sides of the triangle and a parallel to the base through the vertex.

4. Given the sum of the squares on two segments and their ratio, construct the segments.

5. Given the difference of the squares on two segments and their ratio, construct the segments.

6. Given the base and ratio of the sides of a triangle, construct it when any of the following data is given:

   (a) the area;

   (b) the difference on the squares of the sides;

   (c) the sum of the squares on the sides;

   (d) the vertical angle;

   (e) the difference of the base angles.

Proposition 6.11. *PROPORTIONAL SEGMENTS I.*

Given two segments, we wish to find a third proportional segment.

Proof. Construct $\overline{JK}$ and $\overline{LM}$. We wish to construct a segment $\overline{X}$ such that $\overline{JK} : \overline{LM} = \overline{LM} : \overline{X}$.

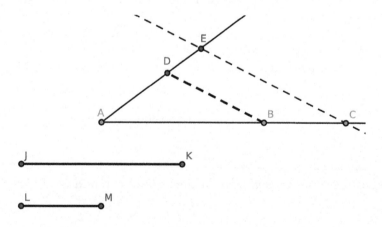

Figure 6.2.13: [6.11]

Construct $\overrightarrow{AC}$ and $\overrightarrow{AE}$ at an arbitrary acute angle. Cut off $\overline{AB} = \overline{JK}$, $\overline{BC} = \overline{LM}$, and $\overline{AD} = \overline{LM}$. Construct $\overline{BD}$ such that $\overline{CE} \parallel \overline{BD}$. We claim that $\overline{DE}$ is the required third proportional segment.

In $\triangle CAE$, $BD \parallel CE$; by [6.2]. $\overline{AB} : \overline{BC} = \overline{AD} : \overline{DE}$. But $\overline{AB} = \overline{JK}$ and $\overline{BC} = \overline{LM} = \overline{AD}$. Hence $\overline{JK} : \overline{LM} = \overline{LM} : \overline{DE}$, which completes the construction. □

6.2. PROPOSITIONS FROM BOOK VI

Remark. Another solution can be inferred from [6.8]. If $\overline{AD}$ and $\overline{DC}$ in that proposition are respectively equal to $\overline{JK}$ and $\overline{LM}$, then $\overline{DB}$ is the third proportional.

Corollary. *6.11.1 Algebraically, this problem can be written as*

$$\frac{a}{b} = \frac{b}{x} \Rightarrow x = \frac{b^2}{a}$$

where a and b are positive real numbers.

Exercises.

1. Suppose that $\triangle AOQ$ is a triangle where $\overline{AQ} > \overline{AO}$. If we cut $\overline{AB} = \overline{AO}$, construct $\overline{BB'} \parallel \overline{AO}$, cut $\overline{BC} = \overline{BB'}$, and so on, prove that the series of segments $\overline{AB}, \overline{BC}, \overline{CD}$, etc., are in continual proportion.

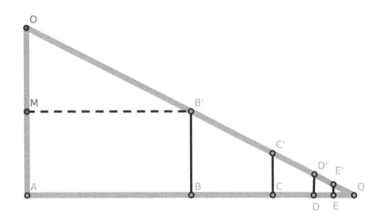

Figure 6.2.14: [6.2, #1]

2. In the above Figure, prove that $(\overline{AB} - \overline{BC}) : \overline{AB} = \overline{AB} : \overline{AQ}$. (Hint: This is evident by constructing $\overline{MB'} \parallel \overline{AQ}$.)

Proposition 6.12. *PROPORTIONAL SEGMENTS II.*

We wish to find a fourth proportional to three given segments.

Proof. Construct $\overline{AK}$, $\overline{BM}$, and $\overline{CP}$. We wish to construct a fourth segment proportional to these three segments; specifically, we wish to construct $\overline{X}$ such that $\overline{AK} : \overline{BM} = \overline{CP} : \overline{X}$.

Construct $\overrightarrow{DE}$ and $\overrightarrow{DF}$ at an arbitrary acute angle. Also construct $\overline{DG} = \overline{AK}$, $\overline{GE} = \overline{BM}$, and $\overline{DH} = \overline{CP}$. Construct $\overline{GH}$ such that $\overline{EF} \parallel \overline{GH}$ [1.31]. We claim that $\overline{HF}$ is the required fourth proportional segment.

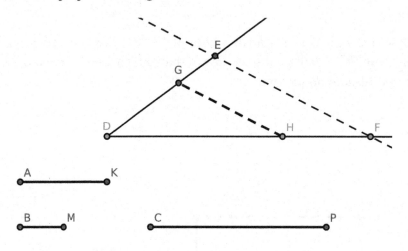

Figure 6.2.15: [6.12]

In $\triangle DEF$, $\overline{GH} \parallel \overline{EF}$, and so $\overline{DG} : \overline{GE} = \overline{DH} : \overline{HF}$ [6.2]. But the above equalities give us $\overline{AK} : \overline{BM} = \overline{CP} : \overline{HF}$. Hence, $\overline{HF}$ is the fourth proportional to $\overline{AK}$, $\overline{BM}$, and $\overline{CP}$, completing the construction. □

Corollary. *6.12.1 Algebraically, this problem can be written as*

$$\frac{a}{b} = \frac{c}{x} \Rightarrow x = \frac{bc}{a}$$

where *a, b,* and *c are positive real numbers. From this equation, it is possible to infer* [Cor. 6.11.1] *where* $c = b$.

6.2. PROPOSITIONS FROM BOOK VI

Proposition 6.13. *PROPORTIONAL SEGMENTS III.*

We wish to find a mean proportional between two given segments.

Proof. Construct $\overline{EF}$ and $\overline{GH}$. We wish to construct a mean proportional between these segments; specifically, we wish to construct $\overline{X}$ such that $\overline{EF} : \overline{X} = \overline{X} : \overline{GH}$.

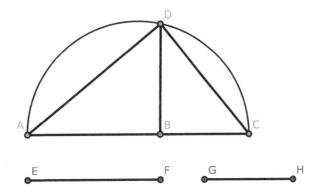

Figure 6.2.16: [6.13]

Construct segment $\overline{AC}$ such that $\overline{AC} = \overline{EF} \oplus \overline{GH}$ where $\overline{AB} = \overline{EF}$ and $\overline{BC} = \overline{GH}$.

On $\overline{AC}$, construct semicircle ADC. Also construct $\overline{BD} \perp \overline{AC}$ which intersects the semicircle at D. We claim that $\overline{BD}$ is the required mean proportional.

Construct $\overline{AD}$ and $\overline{DC}$. Since ADC is a semicircle, $\angle ADC$ is right [3.31]. Since $\triangle ADC$ is a right triangle and $\overline{BD}$ is a perpendicular from the right angle on the hypotenuse, $\overline{BD}$ is a mean proportional between $\overline{AB}$ and $\overline{BC}$ [6.8, Cor. 1]. Thus $\overline{EF} : \overline{BD} = \overline{BD} : \overline{GH}$.

This completes the construction. □

Corollary. 6.13.1. *Algebraically, we have*

$$\frac{a}{x} = \frac{x}{b} \Rightarrow x = \sqrt{ab}$$

where a and b are positive real numbers.

Exercises.

1. If through any point within a circle a chord is constructed which is bisected at that point, prove that its half is a mean proportional between the segments of any other chord passing through the same point.

2. Prove that the tangent to a circle from any external point is a mean proportional between the segments of any secant passing through the same point.

3. If through the midpoint C of any arc of a circle, a secant is constructed cutting the chord of the arc at D and the circle again at E, prove that the chord of half the arc is a mean proportional between $\overline{CD}$ and $\overline{CE}$.

4. If a circle is constructed touching another circle internally and with two parallel chords, prove that the perpendicular from the center of the former on the diameter of the latter, which bisects the chords, is a mean proportional between the two extremes of the three segments into which the diameter is divided by the chords.

5. If a circle is constructed touching a semicircle and its diameter, prove that the diameter of the circle is a harmonic mean between the segments into which the diameter of the semicircle is divided at the point of intersection.

6.2. PROPOSITIONS FROM BOOK VI

Proposition 6.14. *EQUIANGULAR PARALLELOGRAMS.*

Equiangular parallelograms are equal in area if and only if the sides about the equal angles are reciprocally proportional.

Proof. Construct equiangular parallelograms $\square HACB$ and $\square CGDE$. We wish to prove that $\square HACB = \square CGDE$ if and only if $\overline{AC} : \overline{CE} = \overline{GC} : \overline{CB}$.

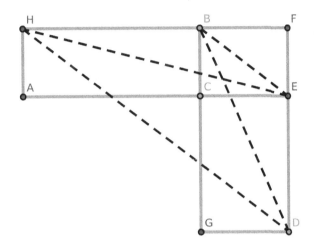

Figure 6.2.17: [6.14]

Suppose that $\square HACB = \square CGDE$; we claim that $\overline{AC} : \overline{CE} = \overline{GC} : \overline{CB}$.

Place $\square HACB$ and $\square CGDE$ so that $\overline{AE} = \overline{AC} \oplus \overline{CE}$ and the equal angles $\angle ACB$ and $\angle ECG$ stand vertically opposite each other. Notice that

$$\angle ACB + \angle BCE = \angle ECG + \angle BCE = \text{two right angles}$$

since $\angle ACB + \angle BCE$ equals two right angles [1.13]. By [1.14], $\overline{BC} \oplus \overline{CG} = \overline{BG}$. Construct $\square BCEF$. Since $\square HACB = \square CGDE$,

$$\begin{aligned}
\overline{AC} : \overline{CE} &= \square HACB : \square BCEF & [6.1] \\
\square HACB : \square BCEF &= \square CGDE : \square BCEF & \text{(hypothesis)} \\
\square CGDE : \square BCEF &= \overline{GC} : \overline{CB} & [6.1]
\end{aligned}$$

Therefore, $\overline{AC} : \overline{CE} = \overline{GC} : \overline{CB}$, which proves our first claim.

Now suppose that $\overline{AC} : \overline{CE} = \overline{GC} : \overline{CB}$. We claim that $\square HACB = \square CGDE$.

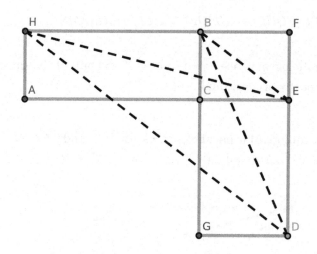

Figure 6.2.18: [6.14]

By [6.1],

$$\overline{AC} : \overline{CE} = \square HACB : \square BCEF$$
$$\overline{GC} : \overline{CB} = \square CGDE : \square BCEF$$

Since $\overline{AC} : \overline{CE} = \overline{GC} : \overline{CB}$ by hypothesis,

$$\square HACB : \square BCEF = \square CGDE : \square BCEF$$

By [5.9], $\square HACB = \square CGDE$, which proves our second and final claim. □

An alternative proof:

Proof. Suppose that $\square HACB = \square CGDE$; we claim that $\overline{AC} : \overline{CE} = \overline{GC} : \overline{CB}$.

Construct $\overline{HE}$, $\overline{BE}$, $\overline{HD}$, and $\overline{BD}$. The area of the parallelogram $\square HACB = 2 \cdot \triangle HBE$, and the area of the parallelogram $\square CGDE = 2 \cdot \triangle BDE$. Therefore $\triangle HBE = \triangle BDE$, and by [1.39.], $\overline{HD} \parallel \overline{BE}$. Hence $\overline{HB} : \overline{BF} = \overline{DE} : \overline{EF}$; that is, $\overline{AC} : \overline{CE} = \overline{GC} : \overline{CB}$.

Part two may be proved by reversing the above, which completes the proof. □

6.2. PROPOSITIONS FROM BOOK VI

Proposition 6.15. *TRIANGLES WITH EQUAL AREAS.*

Construct two triangles which share an equal angle. These triangles have equal area if and only if their sides about the equal angles are reciprocally proportional.

Proof. Construct $\triangle ACB$ and $\triangle DCE$ where $\angle BCA = \angle DCE$. We claim that $\triangle ACB = \triangle DCE$ if and only if $\overline{AC} : \overline{CD} = \overline{EC} : \overline{CB}$.

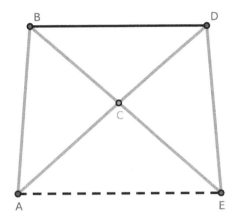

Figure 6.2.19: [6.15]

Suppose that $\triangle ACB = \triangle DCE$; we claim that $\overline{AC} : \overline{CD} = \overline{EC} : \overline{CB}$.

Place $\angle BCA$ and $\angle DCE$ to stand vertically opposite so that $\overline{AD} = \overline{AC} \oplus \overline{CD}$; as in the proof to [6.14], we find that $\overline{BE} = \overline{BC} \oplus \overline{CE}$. Construct $\overline{BD}$.

Since $\triangle ACB = \triangle DCE$,

$$\begin{aligned} \triangle ACB : \triangle BCD &= \triangle DCE : \triangle BCD \\ \triangle ACB : \triangle BCD &= \overline{AC} : \overline{CD} \qquad [6.1] \\ \triangle DCE : \triangle BCD &= \overline{EC} : \overline{CB} \qquad [6.1] \end{aligned}$$

Therefore, $\overline{AC} : \overline{CD} = \overline{EC} : \overline{CB}$, which proves our first claim.

Now suppose that $\overline{AC} : \overline{CD} = \overline{EC} : \overline{CB}$; we claim that $\triangle ACB = \triangle DCE$.

Using the same construction, we have

$$\begin{aligned} \overline{AC} : \overline{CD} &= \overline{EC} : \overline{CB} & \text{(hypothesis)} \\ \overline{AC} : \overline{CD} &= \triangle ACB : \triangle BCD & [6.1] \\ \overline{EC} : \overline{CB} &= \triangle DCE : \triangle BCD & [6.1] \end{aligned}$$

Therefore $\triangle ACB : \triangle BCD = \triangle DCE : \triangle BCD$, and so $\triangle ACB = \triangle DCE$ [5.9]. This proves our second and final claim. $\square$

Remark. [6.15] might have been appended as a corollary to [6.14] since the triangles are the halves of equiangular parallelograms; it may also be proven by constructing $\overline{AE}$ and showing that it is parallel to $\overline{BD}$.

Proposition 6.16. *PROPORTIONAL RECTANGLES.*

Four segments are proportional if and only if the rectangle contained by the extremes (i.e., the largest and the smallest segments) equals the rectangle contained by the means (i.e., the remaining segments).

Proof. We claim that $\overline{AB} : \overline{CD} = \overline{LM} : \overline{NP}$ if and only if $\overline{AB} \cdot \overline{NP} = \overline{CD} \cdot \overline{LM}$. Let $\overline{AB} = x$, $\overline{CD} = y$, $\overline{LM} = u$, and $\overline{NP} = v$. Then

$$\frac{x}{y} = \frac{u}{v} \iff xv = yu$$

which completes the proof. □

A geometric proof:

Proof. Place the four segments in a concurrent position so that the extremes form one continuous segment and the means form a second continuous segment.

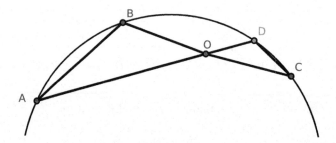

Figure 6.2.20: [6.16], Alternative proof

Place the four segments in the order $\overline{AO}$, $\overline{BO}$, $\overline{OD}$, and $\overline{OC}$. Construct $\overline{AB}$ and $\overline{CD}$. Because $\overline{AO} : \overline{OB} = \overline{OD} : \overline{OC}$ and $\angle AOB = \angle DOC$, the triangles $\triangle AOB$ and $\triangle COD$ are equiangular. Thus, the four points A, B, C, and D are concyclic; by [3.35], $\overline{AO} \cdot \overline{OC} = \overline{BO} \cdot \overline{OD}$. □

6.2. PROPOSITIONS FROM BOOK VI

Proposition 6.17. *LINES AND RECTANGLES.*

Three segments are proportional if and only if the rectangle contained by the first and fourth segments is equal in area to the area of the square of the mean.

Proof. Construct segments $\overline{AB}, \overline{CD}$, and $\overline{GH}$ so that they are proportional: $\overline{AB} : \overline{CD} = \overline{CD} : \overline{GH}$. We claim that $\overline{AB} \cdot \overline{GH} = (\overline{CD})^2$.

Let $\overline{AB} = x$, $\overline{CD} = y$, and $\overline{GH} = z$. Then

$$\frac{x}{y} = \frac{y}{z} \Rightarrow xz = y^2$$

This proves our claim. □

Remark. This proposition may also be inferred as a corollary to [6.16] by setting $y = u$.

Exercises.

1. If a segment $\overline{CD}$ bisects the vertical angle at C of an arbitrary triangle, $\triangle ACB$, prove that its square added to the rectangle $\overline{AD} \cdot \overline{DB}$ contained by the segments of the base is equal in area to the rectangle contained by the sides.

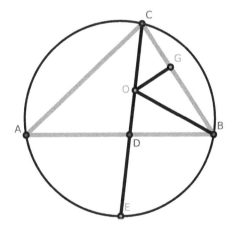

Figure 6.2.21: [6.17, #1]

Hint: Construct a circle about the triangle, and extend $\overline{CD}$ to intersect the circumference at E. Then show that $\triangle ACB$ and $\triangle ECB$ are equiangular. By [6.4], $\overline{AC} : \overline{CD} = \overline{CE} : \overline{CB}$, and by [3.35]),

$$\overline{AC} \cdot \overline{CB} = \overline{CE} \cdot \overline{CD}$$
$$= (\overline{CD})^2 + \overline{CD} \cdot \overline{DE}$$
$$= (\overline{CD})^2 + \overline{AD} \cdot \overline{DB}$$

2. If $\overleftrightarrow{CD'}$ bisects the external vertical angle of an arbitrary triangle, $\triangle ACB$, prove that its square subtracted from the rectangle $\overline{AD'} \cdot \overline{D'B}$ is equal in area to $\overline{AC} \cdot \overline{CB}$.

3. If a circle passing through the angle at point A of a parallelogram $\square ABCD$ intersects the two sides $\overline{AB}, \overline{AD}$ again at the points E, G and the diagonal AC again at F, prove that $\overline{AB} \cdot \overline{AE} + \overline{AD} \cdot \overline{AG} = \overline{AC} \cdot \overline{AF}$.

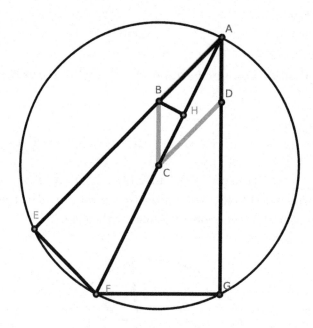

Figure 6.2.22: [6.17, #3]

Hint: construct $\overline{EF}, \overline{FG}$, and $\angle ABH = \angle AFE$. Then the triangles $\triangle ABH$ and $\triangle AFE$ are equiangular: it follows that $\overline{AB} : \overline{AH} = \overline{AF} : \overline{AE}$, and so $\overline{AB} \cdot \overline{AE} = \overline{AF} \cdot \overline{AH}$.

Again, it is clear that the triangles $\triangle BCH$ and $\triangle GAF$ are equiangular, and therefore $\overline{BC} : \overline{CH} = \overline{AF} : \overline{AG}$, and so $\overline{BC} \cdot \overline{AG} = \overline{AF} \cdot \overline{CH}$, or $\overline{AD} \cdot \overline{AG} = \overline{AF} \cdot \overline{CH}$. But since $\overline{AB} \cdot \overline{AE} = \overline{AF} \cdot \overline{AH}$, we find that $\overline{AD} \cdot \overline{AG} + \overline{AB} \cdot \overline{AE} = \overline{AF} \cdot \overline{CH}$.

4. If $\overline{DE}, \overline{DF}$ are parallels to the sides of $\triangle ABC$ from any point D at the base, prove that $\overline{AB} \cdot \overline{AE} + \overline{AC} \cdot \overline{AF} = (\overline{AD})^2 + \overline{BD} \cdot \overline{DC}$.

5. If through a point O within a triangle $\triangle ABC$ parallels $\overline{EF}, \overline{GH}, \overline{IK}$ are constructed to the sides, prove that the sum of the areas of the rectangles constructed by their

segments is equal to the area of the rectangle contained by the segments of any chord of the circumscribing circle passing through O.

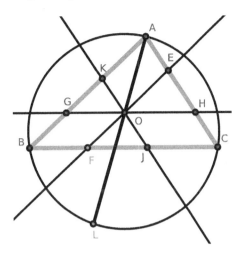

Figure 6.2.23: [6.17, #5]

Hint: notice that

$$\overline{AO} \cdot \overline{AL} = \overline{AB} \cdot \overline{AK} + \overline{AC} \cdot \overline{AE}$$

and

$$(\overline{AO})^2 = \overline{AG} \cdot \overline{AK} + \overline{AH} \cdot \overline{AE} - \overline{GO} \cdot \overline{OH}$$

Hence,

$$\overline{AO} \cdot \overline{OL} = \overline{BG} \cdot \overline{AK} + \overline{CH} \cdot \overline{AE} + \overline{GO} \cdot \overline{OH}$$

or

$$\overline{AO} \cdot \overline{OL} = \overline{EO} \cdot \overline{OF} + \overline{IO} \cdot \overline{OK} + \overline{GO} \cdot \overline{OH}$$

6. Prove that the rectangle contained by the side of an inscribed square standing on the base of a triangle and the sum of the base and altitude equals twice the area of the triangle.

7. Prove that the rectangle contained by the side of an escribed square standing on the base of a triangle and the difference between the base and altitude equals twice the area of the triangle.

8. If from any point P in the circumference of a circle a perpendicular is drawn to any chord, its square is equal in area to the rectangle contained by the perpendiculars from the extremities of the chord on the tangent at P.

9. If O is the point of intersection of the diagonals of a cyclic quadrilateral $ABCD$, prove that the four rectangles $\overline{AB} \cdot \overline{BC}$, $\overline{BD} \cdot \overline{CD}$, $\overline{CD} \cdot \overline{DA}$, $\overline{DA} \cdot \overline{AB}$ are proportional to the four segments $\overline{BO}, \overline{CO}, \overline{DO}, \overline{AO}$.

10. *PTOLEMY'S THEOREM.* The sum of the areas of the rectangles of the opposite sides of a cyclic quadrilateral $ABCD$ equals the area to the rectangle contained by its diagonals.

Hint: construct $\angle DAO = \angle CAB$. Then $\triangle DAO$ and $\triangle CAB$ are equiangular; therefore $\overline{AD} : \overline{DO} = \overline{AC} : \overline{CB}$ and so $\overline{AD} \cdot \overline{BC} = \overline{AC} \cdot \overline{DO}$. Again, the triangles $\triangle DAC$ and $\triangle OAB$ are equiangular, and $\overline{CD} : \overline{AC} = \overline{BO} : \overline{AB}$, or $\overline{AC} \cdot \overline{CD} = \overline{AC} \cdot BO$.

Hence $\overline{AD} \cdot \overline{BC} + \overline{AB} \cdot \overline{CD} = \overline{AC} \cdot \overline{BD}$.

11. If the quadrilateral $ABCD$ is not cyclic, prove that the three rectangles $AB \cdot CD$, $BC \cdot AD$, $AC \cdot BD$ are proportional to the three sides of a triangle which has an angle equal to the sum of a pair of opposite angles of the quadrilateral.

12. Prove by using [6.11] that if perpendiculars fall on the sides and diagonals of a cyclic quadrilateral from any point on the circumference of the circumscribed circle that the rectangle contained by the perpendiculars on the diagonals equals the area of the rectangle contained by the perpendiculars on either pair of opposite sides.

13. If $\overline{AB}$ is the diameter of a semicircle, and $\overline{PA}$, $\overline{PB}$ are chords from any point P in the circumference, and if a perpendicular to $\overline{AB}$ from any point C intersects $\overline{PA}$, $\overline{PB}$ at D and E and the semicircle at F, prove that $\overline{CF}$ is a mean proportional between $\overline{CD}$ and $\overline{CE}$.

6.2. PROPOSITIONS FROM BOOK VI

Proposition 6.18. *CONSTRUCTION OF A SIMILAR POLYGON.*

We may construct a polygon that is similar and similarly placed to a given polygon on a given segment.

Proof. Construct polygon $CDEFG$ and segment AB. We wish to construct a polygon on AB similar to polygon $CDEFG$ and similarly placed.

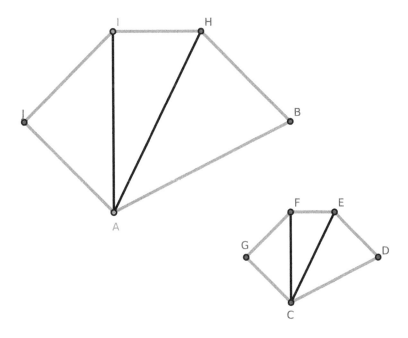

Figure 6.2.24: [6.18]

Construct $\overline{CE}$ and $\overline{CF}$. Also construct $\triangle ABH$ on $\overline{AB}$ such that $\triangle ABH$ is equiangular to $\triangle CDE$ and is similarly placed in regards to $\overline{CD}$; that is, construct $\angle ABH = \angle CDE$ and $\angle BAH = \angle DCE$.

Also construct $\triangle HAI$ equiangular to $\triangle ECF$ and similarly placed. Finally, construct $\triangle IAJ$ equiangular and similarly placed with $\triangle FCG$. We claim that $ABHIJ$ is the required polygon.

By construction, it is evident that the figures are equiangular, and it is only required to prove that the sides about the equal angles are proportional.

Because $\triangle ABH$ is equiangular to $\triangle CDE$, we find that $\overline{AB} : \overline{BH} = \overline{CD} : \overline{DE}$ [6.4]. Hence the sides about the equal angles at points B and D are proportional.

Again from the same triangles, we have $\overline{BH} : \overline{HA} = \overline{DE} : \overline{EC}$, and from the triangles $\triangle IHA$, $\triangle FEC$, we have $\overline{HA} : \overline{HI} = \overline{EC} : \overline{EF}$. Therefore, $\overline{BH} : \overline{HI} = \overline{DE} : \overline{EF}$, or the sides about the equal angles $\angle BHI$, $\angle DEF$ are proportional.

This result follows about the other equal angles, *mutatis mutandis*. By [Def. 6.1] and our placement of each triangle, the proof is complete. □

Remark. In the above construction, the segment $\overline{AB}$ corresponds to $\overline{CD}$, and it is evident that we may take $\overline{AB}$ to correspond to any other side of the given figure $CDEFG$.

Again, in each case, if the figure $ABHIJ$ is turned round the segment $\overline{AB}$ until it falls on the other side, it will still be similar to the figure $CDEFG$. Hence on a given segment $\overline{AB}$, there can be constructed two figures each similar to a given figure $CDEFG$ and having the given segment $\overline{AB}$ correspond to any given side $\overline{CD}$ of the given figure.

The first of the figures thus constructed is said to be directly similar, and the second is said to be inversely similar to the given figure.

Corollary. *6.18.1. Twice as many polygons may be constructed on $\overline{AB}$ similar to a given polygon $CDEFG$ as that figure has sides.*

Corollary. *6.18.2. If the figure $ABHIJ$ is applied to $CDEFG$ so that the point A coincides with C and that the segment $\overline{AB}$ is placed along $\overline{CD}$, then the points H, I, J will be respectively on the segments $\overline{CE}$, $\overline{CF}, \overline{CG}$. Also, the sides $\overline{BH}, \overline{HI}, \overline{IJ}$ of the one polygon will be respectively parallel to their corresponding sides $\overline{DE}, \overline{EF}, \overline{FG}$ of the other.*

Corollary. *6.18.3. If segments constructed from any point O in the plane of a figure to all its vertices are divided in the same ratio, the segments joining the points of division will form a new figure similar to and having every side parallel to the corresponding side of the original.*

Exercises.

1. Prove [Cor. 6.18.1].

2. Prove [Cor. 6.18.2].

3. Prove [Cor. 6.18.3].

6.2. PROPOSITIONS FROM BOOK VI

Proposition 6.19. *RATIOS OF SIMILAR TRIANGLES.*

The areas of similar triangles have a ratio equal to the square of the ratio of the triangles' corresponding sides.

Proof. Construct $\triangle ABC$ and $\triangle DEF$ such that $\triangle ABC \sim \triangle DEF$ (where $\angle ABC = \angle DEF$) and $\overline{AB} : \overline{BC} = \overline{DE} : \overline{EF}$. We claim that $\triangle ABC : \triangle DEF = \left(\overline{BC}\right)^2 : \left(\overline{EF}\right)^2$.

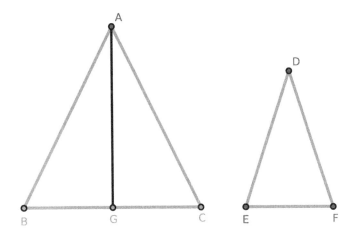

Figure 6.2.25: [6.19]

Using [6.11], construct $\overline{AG}$ such that $\overline{BC} : \overline{EF} = \overline{EF} : \overline{BG}$.

Since $\overline{AB} : \overline{BC} = \overline{DE} : \overline{EF}$ by hypothesis, $\overline{AB} : \overline{DE} = \overline{BC} : \overline{EF}$ by [5.16]. Taken together and applying [5.11], $\overline{AB} : \overline{DE} = \overline{EF} : \overline{BG}$.

Consider $\triangle ABG$ and $\triangle DEF$: the sides about equal angles are reciprocally proportional. By [6.15], $\triangle ABG = \triangle DEF$.

Since $\overline{BC} : \overline{EF} = \overline{EF} : \overline{BG}$, by [Def. 5.9], we have $\left(\overline{BC}\right)^2 : \left(\overline{EF}\right)^2 = \overline{BC} : \overline{BG}$.

By [6.1], we also have $\overline{BC} : \overline{BG} = \triangle ABC : \triangle ABG$, and applying the above, we obtain $\triangle ABC : \triangle ABG = \left(\overline{BC}\right)^2 : \left(\overline{EF}\right)^2$.

Since $\triangle ABG = \triangle DEF$, we have $\triangle ABC : \triangle DEF = \left(\overline{BC}\right)^2 : \left(\overline{EF}\right)^2$, which proves our claim. □

Remark. [6.19] is the first of Euclid's Proposition in which [Def. 5.9], the duplicate ratio, is employed.

An alternate proof:

Proof. Suppose that $\triangle ABC \sim \triangle DEF$. On $\overline{AB}$ and $\overline{DE}$, construct squares $\square AGHB$ and $\square DLME$, respectively. Through points C and F construct segments parallel and respectively equal to $\overline{AB}$ and $\overline{DE}$. Extend $\overline{AG}$, $\overline{BH}$, $\overline{DL}$, and $\overline{EM}$ to points J, I, O, and N, respectively; this constructs rectangles $\square JABI$ and $\square ODEN$.

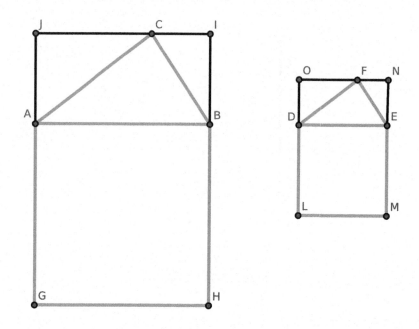

Figure 6.2.26: [6.19] (Casey's proof)

Clearly, $\triangle JAC$ and $\triangle ODF$ are equiangular. By [6.4], $\overline{JA} : \overline{AC} = \overline{OD} : \overline{DF}$ and $\overline{AC} : \overline{AB} = \overline{DF} : \overline{DE}$; thus, $\overline{JA} : \overline{AB} = \overline{OD} : \overline{DE}$. Since $\overline{AB} = \overline{AG}$ and $\overline{DE} = \overline{DL}$ by construction, $\overline{JA} : \overline{AG} = \overline{OD} : \overline{DL}$. By [6.1], $\overline{JA} : \overline{AG} = \square JABI : \square AGHB$ and $\overline{OD} : \overline{DL} = \square ODEN : \square DLME$. Hence

$$\square JABI : \square AGHB = \square ODEN : \square DLME$$

By [5.16],

$$\begin{aligned} \square JABI : \square ODEN &= \square AGHB : \square DLME \\ \tfrac{1}{2} \cdot \square JABI : \tfrac{1}{2} \cdot \square ODEN &= (\overline{AB})^2 : (\overline{DE})^2 \\ \triangle ABC : \triangle DEF &= (\overline{AB})^2 : (\overline{DE})^2 \end{aligned}$$

□

Corollary. *6.19.1 If three segments are proportional, then the first is to the third as the figure described on the first is to that which is similar and similarly described on the second.*[4]

Exercises.

1. If one of two similar triangles has a side that is 50% longer than the corresponding sides of the other, determine the ratio of their areas.

2. When the inscribed and circumscribed regular polygons of any common number of sides to a circle have more than four sides, prove that the difference of their areas is less than the square of the side of the inscribed polygon.

3. Prove [Cor. 6.19.1].

---

[4]https://proofwiki.org/wiki/Ratio_of_Areas_of_Similar_Triangles/Porism

Proposition 6.20. *DIVISION OF SIMILAR POLYGONS.*

Similar polygons may be divided such that:

(1) they divide into the same number of similar triangles,

(2) corresponding triangles have the same ratio to one another as the polygons have to each other,

(3) the polygons have a duplicate ratio of their corresponding sides.

Proof. Construct polygons $ABHIJ$ and $CDEFG$ such that $ABHIJ \sim CDEFG$ and sides $\overline{AB}$ and $\overline{CD}$ correspond to each other. Also construct $\overline{AH}$, $\overline{AI}$, $\overline{CE}$, and $\overline{CF}$. We shall prove each claim separately.

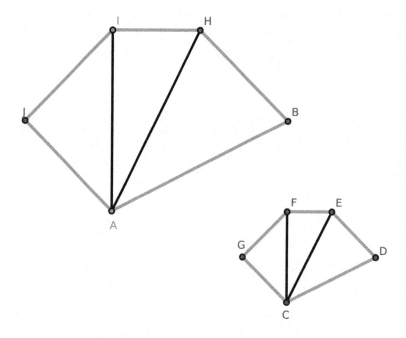

Figure 6.2.27: [6.20]

Claim 1: $ABHIJ$ and $CDEFG$ divide into the same number of similar triangles.

Since $ABHIJ \sim CDEFG$, $ABHIJ$ and $CDEFG$ are equiangular and have proportional sides about their equal angles. It follows that $\angle ABH = \angle CDE$ and $\overline{AB} : \overline{BH} = \overline{CD} : \overline{DE}$. By [6.6], $\triangle ABH$ is equiangular to $\triangle CDE$, and so $\angle BHA = \angle DEC$. Since $\angle BHI = \angle DEF$ by hypothesis, it follows that $\angle AHI = \angle CEF$.

Again, since $ABHIJ \sim CDEFG$, $\overline{IH} : \overline{HB} = \overline{FE} : \overline{ED}$. Since $\triangle ABH$ and $\triangle CDE$ are equiangular, $\triangle ABH \sim \triangle CDE$ and so $\overline{HB} : \overline{HA} = \overline{ED} : \overline{EC}$. It follows that $\overline{IH} : \overline{HA} = \overline{FE} : \overline{EC}$.

6.2. PROPOSITIONS FROM BOOK VI

Since $\angle AHI = \angle CEF$ and $\overline{IH} : \overline{HA} = \overline{FE} : \overline{EC}$, $\triangle AHI \sim \triangle CEF$. Similarly, we may show that all remaining triangles are also equiangular, which proves our claim.

Claim 2: corresponding triangles within $ABHIJ$ and $CDEFG$ have the same ratio to one another as the polygons have to each other.

Since $\triangle ABH \sim \triangle CDE$, by [6.19], $\triangle ABH : \triangle CDE$ is in the duplicate ratio of $\overline{AH} : \overline{CE}$.

Also by [6.19], $\triangle AHI : \triangle CEF$ is in the duplicate ratio of $\overline{AH} : \overline{CE}$.

Hence, $\triangle ABH : \triangle CDE = \triangle AHI : \triangle CEF$ and $\triangle AHI : \triangle CEF = \triangle AIJ : \triangle CFG$, *mutatis mutandis*. Clearly, $\triangle ABH : \triangle CDE = \triangle AIJ : \triangle CFG$. By [5.12],

$$\frac{\triangle ABH}{\triangle CDE} = \frac{\triangle ABH + \triangle AHI + \triangle AIJ}{\triangle CDE + \triangle CEF + \triangle CFG}$$

$$\frac{\triangle ABH}{\triangle CDE} = \frac{ABHIJ}{CDEFG}$$

which proves claim 2.

Claim 3: the polygons have a duplicate ratio of their corresponding sides.

By [6.19], $\triangle ABH : \triangle CDE$ is in the duplicate ratio of $\overline{AB} : \overline{CD}$. Since

$$\frac{\triangle ABH}{\triangle CDE} = \frac{ABHIJ}{CDEFG}$$

$ABHIJ : CDEFG$ is also in the duplicate ratio of $\overline{AB} : \overline{CD}$, which proves our third and final claim. □

Corollary. 6.20.1. *The perimeters of similar polygons are to one another in the ratio of their corresponding sides.*

Corollary. 6.20.2. *As squares are to similar polygons, the duplicate ratio of two segments is equal to the ratio of the squares constructed on them.*

Corollary. 6.20.3. *Similar portions of similar figures have the same ratio to each other as the wholes of the figures.*

Corollary. 6.20.4. *Similar portions of the perimeters of similar figures are to each other in the ratio of the whole perimeters.*

Exercises.

1. If two figures are similar, prove that to each point in the plane of one there will be a corresponding point in the plane of the other.

Let $ABCD$ and $A'B'C'D'$ be the two figures and P a point inside of $ABCD$. Construct $\overline{AP}$ and $\overline{BP}$, and also construct a triangle $\triangle A'P'B'$ on $\overline{A'B'}$ similar to $\triangle APB$. Prove that segments from P' to the vertices of $A'B'C'D'$ are proportional to the lines from P to the vertices of $ABCD$.

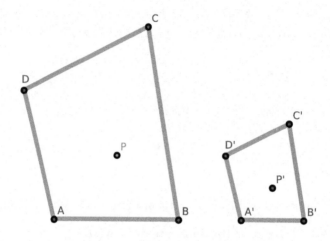

Figure 6.2.28: [6.20] #1

2. If two figures are similar and in the same plane, there is in the plane called a corresponding point with respect to the other (which may be regarded as belonging to either figure).

Let $\overline{AB}$, $\overline{A'B'}$ be two corresponding sides of the figures and C their point of intersection. Through the two triads of points A, A', C and B, B', C construct two circles intersecting again at the point O. Prove that O is the required point. Notice that $\triangle OAB \sim \triangle OAB$ and either may be rotated around O, so that $\overline{AB}$ and $\overline{A'B'}$ will be parallel.

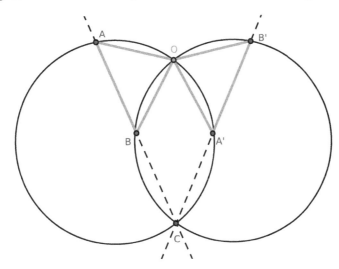

Figure 6.2.29: [6.20] #2 and [Def. 6.9]

3. Prove that two regular polygons of n sides each have n centers of similitude.

4. If any number of similar triangles have their corresponding vertices lying on three given lines, they have a common center of similitude.

5. If two figures are directly similar and have a pair of corresponding sides parallel, every pair of corresponding sides will be parallel.

6. If two figures are homothetic [Def. 6.10], the segments joining corresponding vertices are concurrent, and the point of concurrence is the center of similitude of the figures.

7. If two polygons are directly similar, either may be turned round their center of similitude until they become homothetic, and this may be done in two different ways.

8. Prove that sectors of circles having equal central angles are similar figures.

9. As any two points of two circles may be regarded as corresponding, two circles have in consequence an infinite number of centers of similitude. Their locus is the circle, whose diameter is the line joining the two points for which the two circles are homothetic.

10. The areas of circles are to one another as the squares of their diameters. For they are to one another as the similar elementary triangles into which they are divided, and

these are as the squares of the radii.

11. The circumferences of circles are proportional to their diameters (see [6.20, Cor. 1]).

12. The circumference of sectors having equal central angles are proportional to their radii. Hence if a, a' denote the arcs of two sectors which stand opposite equal angles at the centers, and if r, r' are their radii, then we find that $\frac{a}{r} = \frac{a'}{r'}$.

13. The area of a sector of a circle is equal to half the rectangle contained by the arc of the sector and the radius of the circle.

14. Prove [Cor. 6.20.1].

15. Prove [Cor. 6.20.2].

16. Prove [Cor. 6.20.3].

17. Prove [Cor. 6.20.4].

6.2. PROPOSITIONS FROM BOOK VI

Proposition 6.21. *TRANSITIVITY OF SIMILAR POLYGONS.*

Polygons which are similar to the same figure are similar to one another.

Proof. Construct polygons ABC, DEF, and GHI such that $ABC \sim GHI$ and $DEF \sim GHI$. We claim that $ABC \sim DEF$.

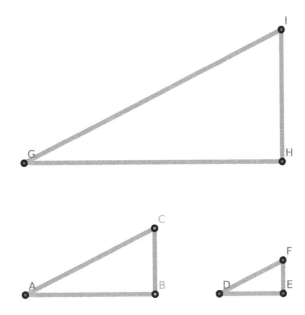

Figure 6.2.30: [6.21] Note that the polygons need not be triangles.

Since $ABC \sim GHI$, they are equiangular and have the sides about their equal angles proportional. Similarly, DEF and GHI are equiangular and have the sides about their equal angles proportional.

Hence ABC and DEF are equiangular and have the sides about their equal angles proportional; or, $ABC \sim DEF$. This completes the proof. □

Remark. Our proof did not use any properties of triangles that are absent in an arbitrary n−sided polygon.

Corollary. *6.21.1. Two similar polygons which are homothetic to a third are homothetic to one another.*

Exercises.

1. If three similar polygons are respectively homothetic, then their three centers of similitudes are collinear.

Proposition 6.22. *PROPORTIONALITY OF FOUR SEGMENTS TO THE POLYGONS CONSTRUCTED UPON THEM.*

Four segments are proportional if and only if the rectilinear figures similar and similarly described upon them are also proportional.

Proof. Suppose that $\overline{AB} : \overline{CD} = \overline{EF} : \overline{GH}$. Construct similar polygons $\triangle ABK$ and $\triangle CDL$ on $\overline{AB}$ and $\overline{CD}$ as well as $\square MEFI$ and $\square NGHJ$ on $\overline{EF}$ and $\overline{GH}$.

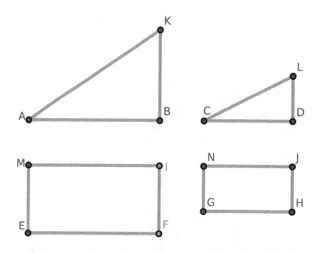

Figure 6.2.31: [6.22]

Suppose $\overline{AB} : \overline{CD} = \overline{EF} : \overline{GH}$. We wish to show that $\triangle ABK : \triangle CDL = \square HEFI : \square NGHJ$. Notice that:

$$\begin{aligned} (\overline{AB})^2 : (\overline{CD})^2 &= (\overline{EF})^2 : (\overline{GH})^2 \\ \triangle ABK : \triangle CDL &= (\overline{AB})^2 : (\overline{CD})^2 \quad [6.20] \\ \square HEFI : \square NGHJ &= (\overline{EF})^2 : (\overline{GH})^2 \quad [6.20] \end{aligned}$$

It follows that $\triangle ABK : \triangle CDL = \square HEFI : \square NGHJ$, which proves our first claim.

Now suppose $\triangle ABK : \triangle CDL = \square HEFI : \square NGHJ$; similarly to the above, we obtain $(\overline{AB})^2 : (\overline{CD})^2 = (\overline{EF})^2 : (\overline{GH})^2$. By [5.22, Cor. 1], $\overline{AB} : \overline{CD} = \overline{EF} : \overline{GH}$, which proves our second and final claim. □

Proposition 6.23. EQUIANGULAR PARALLELOGRAMS.

The areas of equiangular parallelograms have a ratio to each other equal to the ratio of the rectangles contained by their sides about a pair of equal angles.

Proof. Construct equiangular parallelograms $\square HABD$ and $\square BGEC$ where $\angle ABD = \angle GBC$. We claim that $\square HABD : \square BGEC = (\overline{AB} \cdot \overline{BD}) : (\overline{BC} \cdot \overline{BG})$.

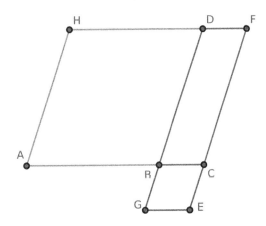

Figure 6.2.32: [6.23]

Let the sides $\overline{AB}$ and $\overline{BC}$ about the equal angles $\angle ABD$ and $\angle CBG$ be placed such that $\overline{AC} = \overline{AB} \oplus \overline{BC}$. As in [6.14], $\overline{GB} \oplus \overline{BD} = \overline{GD}$.

Complete the parallelogram $\square DBCF$. By [6.1],

$$\square HABD : \square DBCF = \overline{AB} : \overline{BC}$$
$$\square DBCF : \square BGEC = \overline{BD} : \overline{BG}$$

It follows that

$$(\square HABD \cdot \square DBCF) : (\square DBCF \cdot \square BGEC) = (\overline{AB} \cdot \overline{BD}) : (\overline{BC} \cdot \overline{BG})$$
$$\square HABD : \square BGEC = (\overline{AB} \cdot \overline{BD}) : (\overline{BC} \cdot \overline{BG})$$

which completes the proof. □

Exercises.

1. Triangles which have one angle of one equal or supplemental to one angle of the other are to one another in the ratio of the rectangles of the sides about those angles.

2. Two quadrilaterals whose diagonals intersect at equal angles are to one another in the ratio of the rectangles of the diagonals.

Proposition 6.24. *SIMILAR PARALLELOGRAMS ABOUT THE DIAGONAL.*

In any parallelogram, the parallelograms about the diagonal are similar both to the whole and to one another.

Proof. Construct $\square ABCD, \overline{GH} \parallel \overline{AB}$ where $\overline{GH} = \overline{AB}$, and $\overline{EK} \parallel \overline{AD}$ where $\overline{EK} = \overline{AD}$. Also construct diagonal $\overline{AC}$. We claim that $\square ABCD \sim \square AEFG$, $\square ABCD \sim \square FHCK$, and $\square AEFG \sim \square FHCK$.

Consider $\triangle ABC$: since $\overleftrightarrow{EF} \parallel \overline{BC}$, by [6.2] $\overline{BE} : \overline{EA} = \overline{CF} : \overline{FA}$.

Consider $\triangle ACD$: since $\overleftrightarrow{FG} \parallel \overline{CD}$, also by [6.2] $\overline{CF} : \overline{FA} = \overline{DG} : \overline{GA}$. It follows that $\overline{BE} : \overline{EA} = \overline{DG} : \overline{GA}$.

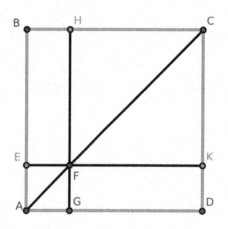

Figure 6.2.33: [6.24]

By [5.18], $\overline{BA} : \overline{EA} = \overline{AD} : \overline{AG}$, and so $\overline{BA} : \overline{AD} = \overline{EA} : \overline{AG}$. Notice that in $\square ABCD$ and $\square AEFG$ the sides about $\angle DAB$ are proportional.

Since $\overleftrightarrow{GF} \parallel \overline{DC}$, $\angle AFG = \angle ACD$. Since $\triangle ADC$ and $\triangle AGF$ share $\angle DAC$ and $\angle ACD$, it follows that $\triangle ADC$ and $\triangle AGF$ are equiangular. Likewise, $\triangle ACB$ and $\triangle AFE$ are equiangular, and so $\square ABCD$ and $\square AEFG$ are also equiangular.

It follows that $\overline{AD} : \overline{DC} = \overline{AG} : \overline{GF}$, $\overline{DC} : \overline{AC} = \overline{GF} : \overline{AF}$, $\overline{AC} : \overline{CB} = \overline{AF} : \overline{FE}$, and $\overline{CB} : \overline{BA} = \overline{FE} : \overline{EA}$. Notice that $\overline{DC} : \overline{CB} = \overline{GF} : \overline{FE}$, or the sides about $\square ABCD$ and $\square AEFG$ are proportional. By [Def. 6.1], $\square ABCD \sim \square AEFG$.

Likewise, $\square ABCD \sim \square FHCK$. By [6.21], $\square AEFG \sim \square FHCK$, which completes the proof. $\square$

Corollary. *6.24.1. Taken in pairs, the parallelograms $\square AEFG$, $\square FHCK$, and $\square ABCD$ are homothetic.*

6.2. PROPOSITIONS FROM BOOK VI

Proposition 6.25. *CONSTRUCTION OF A POLYGON EQUAL IN AREA TO A GIVEN FIGURE AND SIMILAR TO A SECOND GIVEN FIGURE.*

Proof. We wish to construct a polygon equal in area to $ALMN$ but similar to polygon BCD.

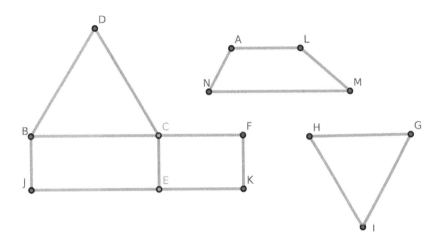

Figure 6.2.34: [6.25] Note that polygon BCD need not be a triangle.

Wlog, on side $\overline{BC}$ of the polygon BCD, construct the rectangle $\square BJEC = \triangle BCD$ [1.44], and on $\overline{CE}$ construct the rectangle $\square CEKF = ALMN$ [1.45].

Construct $\overline{GH}$ such that $\overline{BC} : \overline{GH} = \overline{GH} : \overline{CF}$ [6.13]. On $\overline{GH}$, construct polygon $GHI \sim BCD$ [6.18] where $\overline{BC}$ and $\overline{GH}$ are corresponding sides. We claim that GHI is the required polygon.

Since $\overline{BC} : \overline{GH} = \overline{GH} : \overline{CF}$,
$$\frac{BC}{CF} = \left(\frac{BC}{GH}\right)^2$$

Since $BCD \sim GHI$, by [6.20]
$$\frac{BCD}{GHI} = \left(\frac{BC}{GH}\right)^2$$

We also have that $\overline{BC} : \overline{CF} = \square BJEC : \square CEKF$, and so $\square BJCE : \square CEKF = BCD : GHI$.

But the rectangle $\square BJEC$ is equal in area to the polygon BCD; therefore, $\square CEKF = GHI$. Since $\square CEKF = ALMN$ by construction, it follows that $GHI = ALMN$ where $GHI \sim BCD$ by construction. □

An alternate proof:

Proof. Construct squares $\square EFJK$ and $\square LMNO$ such that $\square EFJK =$ polygon DCB and $\square LMNO =$ polygon $APQS$ [2.14]. By [6.12], construct $\overline{GH}$ such that $\overline{EF} : \overline{LM} = \overline{BC} : \overline{GH}$.

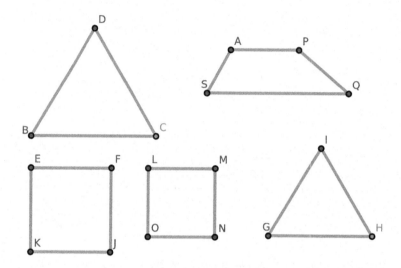

Figure 6.2.35: [6.25], alternate proof. Note that BCD need not be a triangle.

On $\overline{GH}$, construct the polygon GHI similar to the polygon BCD [6.18] such that $\overline{BC}$ and $\overline{GH}$ are corresponding sides. We claim that GHI is the required polygon.

Because $\overline{EF} : \overline{LM} = \overline{BC} : \overline{GH}$ by construction, we find that $EFJK : LMNO = BCD : GHI$ [6.22]. But $EFJK = BCD$ by construction; therefore, $LMNO = GHI$. But $LMNO = APQS$ by construction. Therefore $GHI = APQS$ and is similar to BCD. $\square$

6.2. PROPOSITIONS FROM BOOK VI

Proposition 6.26. *PARALLELOGRAMS ON A COMMON ANGLE.*

If two similar and similarly situated parallelograms have a common angle, then they stand on the same diagonal.

Proof. Construct $\square AEFG$ and $\square ABCD$ such that each are similar and similarly situated where $\angle GAF$ is a common angle. We claim that they stand on the same diagonal, $\overline{AC}$.

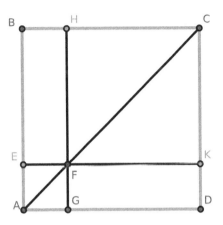

Figure 6.2.36: [6.26]

Construct the diagonals $\overline{AF}$ and $\overline{AC}$. Because $\square AEFG \sim \square ABCD$ by hypothesis, they can be divided into the same number of similar triangles [6.20]. Hence, $\triangle GAF \sim \triangle CAD$, and it follows that $\angle GAF = \angle CAD$.

Hence, $\overline{AC}$ contains point F, and so the parallelograms stand on $\overline{AC}$. $\square$

Remark. [6.26] is the converse of [6.24] and may have been misplaced in an early edition of Euclid. The following would be a simpler statement of result: "If two homothetic parallelograms have a common angle, they stand on the same diagonal."

Proposition 6.27. *INSCRIBING A PARALLELOGRAM IN A TRIANGLE I.*

In a given triangle, we wish to inscribe the parallelogram with maximum area having a common angle with a given triangle.

Proof. Construct $\triangle ABC$ where the given angle is $\angle ABC$. Bisect $\overline{AC}$ at P; through P, construct $\overleftrightarrow{PE} \parallel \overleftrightarrow{BC}$ and $\overleftrightarrow{PF} \parallel \overleftrightarrow{AB}$. We claim that $\square EBFP$ is the required parallelogram.

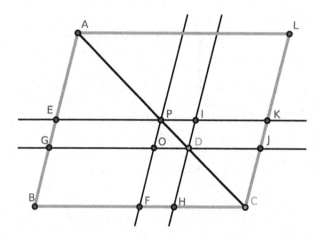

Figure 6.2.37: [6.27]

Construct $\overline{AL} = \overline{BC}$ where $\overline{AL} \parallel \overline{BC}$ as well as $\overline{CL}$. By [1.33], $\overline{CL} \parallel \overline{BA}$ and $\overline{CL} = \overline{BA}$.

Take any point D on $\overline{AC}$ other than P and construct $\overleftrightarrow{DG} \parallel \overline{BC}$ which intersects $\overleftrightarrow{PF}$ at O, $\overline{AB}$ at G, and $\overline{CL}$ at J. Also construct $\overleftrightarrow{DH} \parallel \overline{AB}$ which intersects $\overleftrightarrow{EK}$ at I and $\overline{BC}$ at H.

Since $\overline{AC}$ is bisected at P, $\overline{EK}$ is also bisected in P. By [1.36], $\square EGOP = \square POJK$. Therefore, $\square EGOP > \square IDJK$; but $\square IDJK = \square OFHD$ [1.43], and so $\square EGOP > \square OFHD$.

Add $\square GBFO$ to each, and we find that $\square EBFP > \square GBHD$. Since our choice of D was any point on $\overline{AC}$ other than P, and since $\square EBFP$ contains $\angle ABC$, $\square EBFP$ is the maximum parallelogram which can be inscribed in the triangle $\triangle ABC$ and which contains $\angle ABC$. □

Corollary. 6.27.1. *The maximum parallelogram exceeds the area of any other parallelogram about the same angle in the triangle by the area of the similar parallelogram whose diagonal is the line between the midpoint P of the opposite side and the point D, which is the corner of the other inscribed parallelogram.*

6.2. PROPOSITIONS FROM BOOK VI

Corollary. *6.27.2. The parallelograms inscribed in a triangle and having one angle common with it are proportional to the rectangles contained by the segments of the sides of the triangle made by the opposite corners of the parallelograms.*

Exercises.

1. Prove [Cor. 6.27.1].

2. Prove [Cor. 6.27.2].

CHAPTER 6. APPLICATIONS OF PROPORTIONS

Proposition 6.28. *INSCRIBING A PARALLELOGRAM IN A TRIANGLE II.*

We wish to inscribe in a given triangle a parallelogram equal in area to a given polygon (the area of which is less than or equal to the area of the maximum inscribed polygon constructed in [6.27]) and having an angle in common with the triangle.

Proof. Construct $\triangle ABC$. Bisect $\overline{AC}$ at P, construct $\overline{PF} \perp \overline{BC}$, and construct $\overline{PE} \perp \overline{AB}$. Applying [6.27], construct $\square EBFP$, the maximum inscribed parallelogram within $\triangle ABC$. We wish to inscribe in $\triangle ABC$ a parallelogram equal in area to polygon X (given that area $X \leq \square EBFP$) and which shares $\angle ABC$.

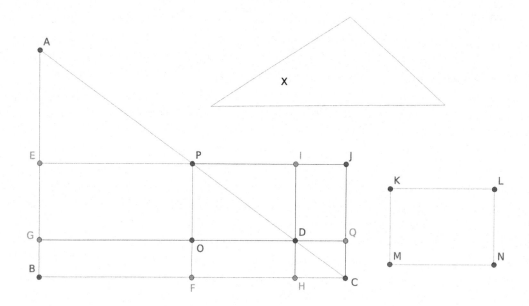

Figure 6.2.38: [6.28]

If area $X = \square EBFP$, the construction is complete.

Otherwise, extend $\overline{EP}$ to $\overline{EJ}$, construct $\overline{CJ} \parallel \overline{PF}$ where $\overline{CJ} = \overline{PF}$ [1.33].

By [6.25], construct $\square KLMN$ such that $\square KLMN = \square PFCJ - X$ and $\square KLMN \sim \square PFCJ$. On $\overline{PJ}$, construct $\overline{PI} = \overline{KL}$. Construct $\overline{IH} = \overline{PF}$ such that $\overline{IH} \parallel \overline{AB}$ and $\overline{IH}$ intersects $\overline{AC}$ at D. Also construct $\overline{DG} \parallel \overline{BC}$. We claim that $\square GBHD$ is the required parallelogram.

Clearly, $\square GBHD$ shares $\angle ABC$ in common with $\triangle ABC$.

Since $\square PFCJ$ and $\square PODI$ stand on the same diagonal, by [6.24] $\square PFCJ \sim \square PODI$. Since $\square PFCJ \sim \square KMNL$ by construction, by [6.21] $\square PODI \sim \square KMNL$. Since $\overline{KL} = \overline{PI}$, $\square PODI = \square KMNL$.

By [Cor. 6.27.1], $\square PODI = \square EBFP - \square GBHD$; by the above, $\square KMNL = \square PFCJ - X$. Hence,

$$\square EBFP - \square GBHD = \square PFCJ - X$$

Since $\overline{BF} = \overline{FC}$ by construction, $\square EBFP = \square PFCJ$. Thus,

$$\square GBHD = X$$

which completes the construction. $\square$

Remark. This proposition geometrically solves the equation $ay - y^2 = C$ where $C =$ the area of polygon X.

Proposition 6.29. *ESCRIBING A PARALLELOGRAM TO A TRIANGLE.*

Given a polygon, a segment, and a parallelogram, we wish to construct a new parallelogram on the given segment whose area is equal to the given polygon. An extension of this new parallelogram will be similar to the given parallelogram.

Proof. We wish to construct a parallelogram ($\square APOI$) equal in area to polygon $CUVWX$ on segment $\overline{AB}$. An extension of $\square APOI$ ($\square BPOQ$) will be similar to $\square DRST$.

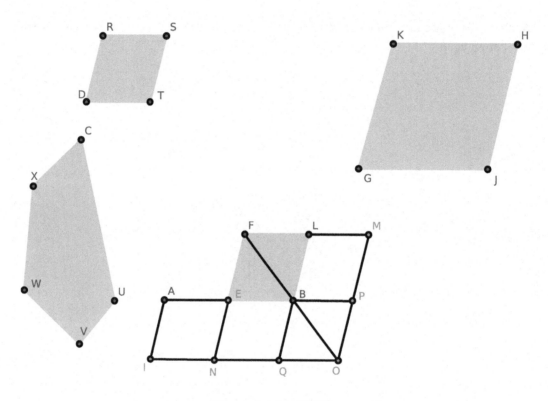

Figure 6.2.39: [6.29]

Bisect $\overline{AB}$ at E. Construct $\square FEBL$ on $\overline{EB}$ such that $\square FEBL \sim \square DRST$ and where $\square FEBL$ is similarly situated to $\square DRST$. By [6.25] construct $\square GKHJ = \square FEBL + CUVWX$ where $\square GKHJ \sim \square DRST$. Let $\overline{KH}$ correspond to $\overline{FL}$ and $\overline{KG}$ correspond to $\overline{FE}$.

Since $\square GKHJ > \square FEBL$, it follows that $\overline{KH} > \overline{FL}$ and $\overline{KG} > \overline{FE}$.

Extend $\overline{FL}$ to $\overline{FM}$ where $\overline{FM} = \overline{KH}$, and extend $\overline{FE}$ to $\overline{FN}$ where $\overline{FN} = \overline{KG}$.

Construct $\square FMON$; by [6.26] $\square FMON = \square GKHJ$ and $\square FMON \sim \square GKHJ$.

But $\square GKHJ \sim \square FEBL$, and so $\square FMON \sim \square FEBL$ [6.21]. By [6.26], $\square FMON$ and $\square FEBL$ stand on the same diameter.

6.2. PROPOSITIONS FROM BOOK VI

Construct diameter $\overline{FO}$ of $\square FMON$. Since $\square GKHJ = \square FEBL + CUVWX$ and $\square GKHJ = \square FMON$, $\square FMON = \square FEBL + CUVWX$. Subtracting the area of $\square FEBL$ from each side of the equation, we obtain gnomon $EBLMON = CUVWX$.

Construct $\square AINE$ where $\overline{AE} = \overline{EB}$. Also construct $\overline{BP} = \overline{LM}$ where $\overline{BP} \parallel \overline{LM}$, and $\overline{BQ} = \overline{PO}$ where $\overline{BQ} \parallel \overline{PO}$. Finally, construct and $\overline{BQ} = \overline{EN}$.

By construction, $\square AENI = \square EBQN$. By [1.43], $\square EBQN = \square LMBP$, and so $\square AENI = \square LMBP$. Hence

$$\begin{aligned} \square AENI + \square EPON &= \square LMBP + \square EPON \\ \square APOI &= EBLMON \\ \square APOI &= CUVWX \end{aligned}$$

Clearly, $\square APOI$ is constructed on $\overline{AB}$, and $\square BPOQ \sim \square DRST$. This completes the construction. $\square$

Remark. This proposition geometrically solves the equation $ax + x^2 = C$ where $C =$ the area of polygon $CUVWX$.

Proposition 6.30. *EXTREME AND MEAN RATIO OF A SEGMENT (aka. THE GOLDEN SECTION)*

A segment may be divided into its "extreme and mean ratio."

Proof. On an arbitrary segment $\overline{AB}$, divide $\overline{AB}$ at E such that $\overline{AB} \cdot \overline{BE} = \left(\overline{AE}\right)^2$ [2.11].

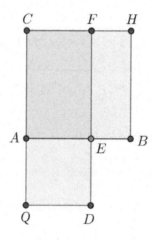

Figure 6.2.40: [6.30] Copyright Prime.mover & Daniel Callahan, licensed under CC SA 3.0

Hence, we obtain

$$\begin{aligned} \overline{AB} \cdot \overline{BE} &= \overline{AE} \cdot \overline{AE} \\ \overline{AB} \cdot \overline{BE} \cdot \frac{1}{\overline{BE} \cdot \overline{AE}} &= \overline{AE} \cdot \overline{AE} \cdot \frac{1}{\overline{BE} \cdot \overline{AE}} \\ \frac{\overline{AB}}{\overline{AE}} &= \frac{\overline{AE}}{\overline{BE}} \end{aligned}$$

or $\overline{AB} : \overline{AE} = \overline{AE} : \overline{BE}$. □

Exercises.

1. If the three sides of a right triangle are in continued proportion, prove that the hypotenuse is divided in extreme and mean ratio by the perpendicular from the right angle on the hypotenuse.

2. In the same case as #1, prove that the greater segment of the hypotenuse is equal to the least side of the triangle.

6.2. PROPOSITIONS FROM BOOK VI

Proposition 6.31. *AREA OF SQUARES ON A RIGHT TRIANGLE.*

If any similar quadrilateral is similarly constructed on the three sides of a right triangle, the quadrilateral on the hypotenuse is equal in area to the sum of the areas of the quadrilaterals constructed on the two other sides.

Proof. Construct $\triangle ABC$. Denote the sides of $\triangle ABC$ by a, b, and c where c is the hypotenuse, and denote the corresponding areas of the similar polygons by α, β, and γ. We claim that $\alpha + \beta = \gamma$.

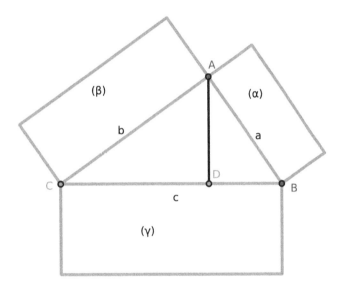

Figure 6.2.41: [6.31]

Because the polygons are similar, by [6.20]

$$\frac{\alpha}{\gamma} = \frac{a^2}{c^2} \text{ and } \frac{\beta}{\gamma} = \frac{b^2}{c^2}$$

It follows that

$$\frac{\alpha + \beta}{\gamma} = \frac{a^2 + b^2}{c^2}$$

But $a^2 + b^2 = c^2$ by [1.47]. Therefore, $\alpha + \beta = \gamma$, which proves our claim. □

Exercise.

1. If semicircles are constructed on supplemental chords of a semicircle, prove that the sum of the areas of the two crescents thus formed is equal to the area of the triangle whose sides are the supplemental chords and the diameter.

Proposition 6.32. *FORMATION OF TRIANGLES.*

If two triangles exist such that a pair of sides in one is proportional to a pair of sides in the other and placed such that their corresponding sides are parallel, then the remaining sides of the triangles form a segment.

Proof. Construct $\triangle ABC$ and $\triangle DCE$ such that $\overline{AB} : \overline{AC} = \overline{DC} : \overline{DE}$, $\overline{AB} \parallel \overline{DC}$, and $\overline{AC} \parallel \overline{DE}$. We claim that $\overline{BC} \oplus \overline{CE} = \overline{BE}$.

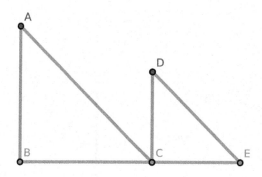

Figure 6.2.42: [6.32]

Since $\overline{AB} \parallel \overline{DC}$ and each intersects $\overline{AC}$, we find that $\angle BAC = \angle ACD$. Similarly, $\angle CDE = \angle ACD$; hence, $\angle BAC = \angle CDE$.

Consider $\triangle ABC$ and $\triangle DCE$: $\angle BAC = \angle CDE$ and $\overline{AB} : \overline{AC} = \overline{DC} : \overline{DE}$. By [6.6], $\triangle ABC$ and $\triangle DCE$ are equiangular, and so $\angle ABC = \angle DCE$.

Furthermore,

$$\angle ACE = \angle ACD + \angle DCE$$
$$= \angle BAC + \angle ABC$$
$$\Rightarrow$$
$$\angle ACE + \angle ACB = \angle BAC + \angle ABC + \angle ACB$$
$$= \text{two right angles [1.32]}$$

Thus, $\overline{BE} = \overline{BC} \oplus \overline{CE}$, which proves our claim. □

6.2. PROPOSITIONS FROM BOOK VI

Proposition 6.33. *RATIOS OF EQUAL TRIANGLES.*

In equal circles, angles at the centers or at the circumferences have the same ratio to one another as the arcs on which they stand.

Proof. Construct $\circ G$ and $\circ H$ with equal radii. Construct $\angle BGC$ at the center of $\circ G$ and $\angle EHF$ at the center of $\circ H$; also construct $\angle BAC$ at the circumference of $\circ G$ and $\angle EDF$ at the circumference of $\circ H$. We wish to show that

$$\text{arc } BC : \text{arc } EF = \angle BGC : \angle EHF = \angle BAC : \angle EDF$$

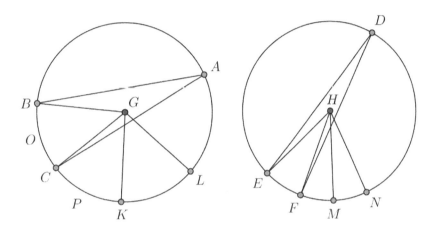

Figure 6.2.43: [6.33] Copyright Prime.mover, licensed under CC SA 3.0

On $\circ G$, construct a finite number n of consecutive arcs CK, KL which are equal in length to arc BC. On $\circ H$, construct n consecutive arcs FM, MN which are equal in length to arc EF.

Construct $\overline{GK}, \overline{GL}, \overline{HM},$ and $\overline{HN}$. Since $BC = CK = KL$, by [3.27] $\angle BGC = \angle CGK = \angle KGL$. It follows that if arc $BL = n \cdot$ arc BC, then $\angle BGL = n \cdot \angle BGC$. Similarly, if arc $EN = n \cdot$ arc EF, then $\angle EHN = n \cdot \angle EHF$.

Since $BL > 0$ and $EN > 0$, $\frac{BL}{EN} = k$ where $k > 0$ is a positive real number. Hence, $BL = k \cdot EN$.

Since
$$\frac{\angle EHN}{\angle EHF} = \frac{\angle BGL}{\angle BGC} = n = \frac{BL}{BC} = \frac{EN}{EF}$$

it follows that
$$\frac{k \cdot EN}{BC} = \frac{k \cdot EN}{k \cdot EF}$$

and so
$$\frac{BC}{EF} = k$$

Similarly, $\angle BGL = k \cdot \angle EHN$, and so $BC : EF = \angle BGC : \angle EHF$. Applying [3.20], we obtain
$$BC : EF = \angle BGC : \angle EHF = \angle BAC : \angle DEF$$

This proves our claim. □

Corollary. *6.33.1 In equal circles, sectors have the same ratio to one another as the arcs on which they stand.*

Exercises.

#1. Prove [Cor. 6.33.1].

6.2. PROPOSITIONS FROM BOOK VI

Exam questions for chapter 6.

1. What is the subject-matter of chapter 6? (Ans. Application of the theory of proportion.)

2. What are similar polygons?

3. What do similar polygons agree in?

4. How many conditions are necessary to define similar triangles?

5. How many conditions are necessary to define similar polygons of more than three sides?

6. When is a polygon said to be given in species?

7. What is a mean proportional between two lines?

8. Define two mean proportionals.

9. What is the altitude of a polygon?

10. If two triangles have equal altitudes, how do their areas vary?

11. How do these areas vary if they have equal bases but unequal altitudes?

12. If both bases and altitudes differ, how do the areas vary?

13. When are two segments divided proportionally?

14. If two triangles have equal areas, prove that their perpendiculars are reciprocally proportional to the bases.

15. What is meant by inversely similar polygons?

16. How many polygons similar to a given polygon of sides can be constructed on a given line?

17. What are homothetic polygons?

18. How do the areas of similar polygons vary?

19. What proposition is [6.19] a special case of?

Exercises for chapter 6.

1. If a transversal meets the sides of a triangle $\triangle ABC$ at the points A', B', C', prove that $\overline{AB'} \cdot \overline{BC'} \cdot \overline{CA'} = -\overline{A'B} \cdot \overline{B'C} \cdot \overline{C'A}$.

2. If D is the midpoint of the base BC of a triangle $\triangle ABC$, E the foot of the perpendicular, L is the point where the bisector of the angle at A meets BC, and H the point of intersection of the inscribed circle with BC, prove that $\overline{DE} \cdot \overline{HL} = \overline{HE} \cdot \overline{HD}$.

3. As in #2, if K is the point of intersection with $\overline{BC}$ of the escribed circle, which touches the other extended sides, prove that $\overline{LH} \cdot \overline{BK} = \overline{BD} \cdot \overline{LE}$.

4. If R, r, r', r'', r''' are the radii of the circumscribed, the inscribed, and the escribed circles of a plane triangle, d, d', d'', d''' the distances of the center of the circumscribed circle from the centers of the others, prove that $R^2 = d^2 + 2Rr = d'^2 - 2Rr'$, etc.

5. As in #4, prove that $12R^2 = d^2 + d'^2 + d''^2 + d'''^2$.

6. If p', p'', p''' denote the altitudes of a triangle, then:

(1) $\frac{1}{p'} + \frac{1}{p''} + \frac{1}{p'''} = \frac{1}{r}$

(2) $\frac{1}{p''} + \frac{1}{p'''} - \frac{1}{p} = \frac{1}{r'}$ (etc.)

(3) $\frac{2}{p} = \frac{1}{r} - \frac{1}{r'}$ (etc.)

(4) $\frac{2}{p'} = \frac{1}{r''} + \frac{1}{r'''}$ (etc.)

7. Suppose that the angle at point A and the area of a triangle $\triangle ABC$ are given in magnitude. If the point A is fixed in position and the point B move along a fixed line or circle, prove that the locus of the point C is a circle.

8. Find the area of a triangle:

(a) in terms of its medians;

(b) in terms of its perpendiculars.

9. If there are three given parallel lines and two fixed points A, B, and if the lines connecting A and B to any variable point in one of the parallels intersects the other parallels at the points C and D, E and F, respectively, prove that $\overline{CF}$ and $\overline{DE}$ each pass through a fixed point.

10. Find a point O in the plane of a triangle $\triangle ABC$ such that the diameters of the three circles about the triangles $\triangle OAB$, $\triangle OBC$, $\triangle OCA$ may be in the ratios of three given segments.

11. Suppose that $ABCD$ is a cyclic quadrilateral, and the segments $\overline{AB}, \overline{AD}$, and the point C are given in position. Find the locus of the point which divides $\overline{BD}$ in a given ratio.

12. If $\overline{CA}, \overline{CB}$ are two tangents to a circle and $\overline{BE} \perp \overline{AD}$ (where $\overline{AD}$ is the the diameter through A), prove that $\overline{CD}$ bisects $\overline{BE}$.

13. If three segments from the vertices of a triangle $\triangle ABC$ to any interior point O meet the opposite sides in the points A', B', C', prove that

$$\frac{OA'}{AA'} + \frac{OB'}{BB'} + \frac{OC'}{CC'} = 1$$

14. If three concurrent segments OA, OB, OC are cut by two transversals in the two systems of points A, B, C; A', B', C', respectively, then prove that

6.2. PROPOSITIONS FROM BOOK VI

$$\frac{AB}{A'B'} \cdot \frac{OC}{OC'} = \frac{BC}{B'C'} \cdot \frac{OA}{OA'} = \frac{CA}{C'A'} \cdot \frac{OB}{OB'}$$

15. Prove that the line joining the midpoints of the diagonals of a quadrilateral circumscribed to a circle:

(a) divides each pair of opposite sides into inversely proportional segments;

(b) is divided by each pair of opposite segments into segments which when measured from the center are proportional to the sides;

(c) is divided by both pairs of opposite sides into segments which when measured from either diagonal have the same ratio to each other.

16. If $\overline{CD}, \overline{CD'}$ are the internal and external bisectors of the angle at C of the triangle $\triangle ACB$, prove that the three rectangles $\overline{AD} \cdot \overline{DB}, \overline{AC} \cdot \overline{CB}, \overline{AD} \cdot \overline{BD}$ are proportional to the squares of $\overline{AD}, \overline{AC}, \overline{AD}$ and are:

(a) in arithmetical progression, if the difference of the base angles is equal to a right angle;

(b) in geometrical progression if one base angle is right;

(c) in harmonic progression if the sum of the base angles is equal to a right angle.

17. If a variable circle touches two fixed circles, the chord of contact passes through a fixed point on the line connecting the centers of the fixed circles.

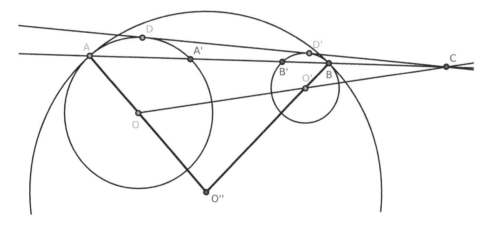

Figure 6.2.44: Ch. 6, #27

Let O, O' be the centers of the two fixed circles where O is the center of the variable circle. Let A, B the points of contact, and let $\overline{AB}$ and $\overline{OO'}$ meet at C, and cut the fixed circles again in the points A', B' respectively.

Construct $\overline{A'O}, \overline{AO}, \overline{BO'}$. Then $\overline{AO}, \overline{BO'}$ meet at O'' [3.11]. Now because $\triangle OAA'$, $\triangle O''AB$ are isosceles, the angles $\angle O''BA = \angle O''AB = \angle OA'A$.

Hence $\overline{OA'} \parallel \overline{O'B}$; therefore $\overline{OC} : \overline{O'C} = \overline{OA'} : \overline{O'B}$ is in a given ratio. Hence, C is a given point.

18. In #17, if $\overline{DD'}$ is the common tangent to the two circles, prove that $(\overline{DD'})^2 = \overline{AB'} \cdot \overline{A'B}$.

19. If R denotes the radius of O'' and ρ, ρ' the radii of O, O', then $(\overline{DD'})^2 : (\overline{AB})^2 = (R \pm \rho)(R \pm \rho') : R^2$ where the choice of sign depends on the nature of the contacts. (This result follows from #18.)

20. Prove that the inscribed and escribed circles of any triangle are all touched by its nine-points circle.

21. If a, b, c, d denote the four sides of a quadrilateral, and D, D' denote the diagonals of a quadrilateral, prove that the sides of the triangle, formed by joining the feet of the perpendiculars from any of its vertices on the sides of the triangle formed by the three remaining points, are proportional to the three rectangles ac, bd, DD'.

22. Prove the converse of Ptolemy's theorem (see [6.17], #10).

23. Construct a circle which:

    (a) passes through a given point, and touches two given circles;

    (b) touches three given circles.

24. Prove that if a variable circle touches two fixed circles, the tangent to it from their center of similitude through which the chord of contact passes is of constant length. (See #17 above.)

25. If segments $\overline{AD}$, $\overline{BD'}$ are extended, prove that they meet at a point on the circumference of O'' and the line $O''P$ is perpendicular to DD'. (See #17 above.)

26. If a segment $\overline{EF}$ divides proportionally two opposite sides of a quadrilateral, and a segment $\overline{GH}$ the other sides, prove that each of these is divided by the other in the same ratio as the sides which determine them.

27. In a given circle, inscribe a triangle such that the triangle whose vertices are the feet of the perpendiculars from the endpoints of the base on the bisector of the vertical angle and the foot of the perpendicular from the vertical angle on the base may be a maximum.

28. In a circle, prove that the point of intersection of the diagonals of any inscribed quadrilateral coincides with the point of intersection of the diagonals of the circumscribed quadrilateral whose sides touch the circle at the vertices of the inscribed quadrilateral.

29. Through two given points construct a circle whose common chord with another given circle may be parallel to a given line, or pass through a given point.

30. If concurrent lines constructed from the angles of a polygon of an odd number of sides divide the opposite sides each into two segments, prove that the product of one set of alternate segments is equal in area to the product of the other set.

31. If a triangle is constructed about a circle, prove that the lines from the points of contact of its sides with the circle to the opposite vertices are concurrent.

32. If a triangle is inscribed in a circle, prove that the tangents to the circle at its three vertices meet the three opposite sides at three collinear points.

33. Prove that the external bisectors of the angles of a triangle meet the opposite sides in three collinear points.

34. Construct a circle touching a given line at a given point and cutting a given circle at a given angle.

35. Prove that the center of mean position of the vertices of a regular polygon is the center of figure of the polygon.

36. Prove that the sum of the squares of segments constructed from any system of points A, B, C, D, etc., to any point P exceeds the sum of the squares of segments from the same points to their center of mean position, O, by $n \cdot (\overline{OP})^2$.

37. If a point is taken within a triangle so as to be the center of mean position of the feet of the perpendiculars constructed from it to the sides of the triangle, prove that the sum of the squares of the perpendiculars is a minimum.

38. Construct a quadrilateral being given two opposite angles, the diagonals, and the angle between the diagonals.

39. Construct two points, C, D in the circumference of a given circle are on the same side of a given diameter. Find a point P in the circumference at the other side of the given diameter, $\overline{AB}$, such that $\overline{PC}$, $\overline{PD}$ may cut $\overline{AB}$ at equal distances from the center.

40. If the sides of any polygon are cut by a transversal, prove that the product of one set of alternate segments is equal to the product of the remaining set.

41. A transversal being constructed cutting the sides of a triangle, prove that the lines from the angles of the triangle to the midpoints of the segments of the transversal intercepted by those angles meet the opposite sides in collinear points.

42. If segments are constructed from any point P to the angles of a triangle, prove that the perpendiculars at P to these segments meet the opposite sides of the triangle at three collinear points.

43. Prove that the rectangle contained by the perpendiculars from the endpoints of the base of a triangle on the internal bisector of the vertical angle is equal to the rectangle contained by the external bisector and the perpendicular from the middle of the base on the internal bisector.

44. State and prove the corresponding theorem for perpendiculars on the external bisector.

45. Suppose that R, R' denote the radii of the circles inscribed in the triangles into which a right triangle is divided by the perpendicular from the right angle on the hypotenuse. If c is the hypotenuse and s is the semi-perimeter, prove that $R^2 + R'^2 = (s-c)^2$.

46. If A, B, C, D are four collinear points, find a point O in the same line with them such that $\overline{OA} \cdot \overline{OD} = \overline{OB} \cdot \overline{OC}$.

47. Suppose the four sides of a cyclic quadrilateral are given; construct it.

48. If a circle touches internally two sides of a triangle, $\overline{CA}$, $\overline{CB}$, and its circumscribed circle, prove that the distance from C to the point of intersection on either side is a fourth proportional to the semi-perimeter, $\overline{CA}$ and $\overline{CB}$.

49. State and prove the corresponding theorem for a circle touching the circumscribed circle externally and two extended sides.

50. Pascal's Theorem: if the opposite sides of an irregular hexagon $ABCDEF$ inscribed in a circle are extended until they meet, the three points of intersection G, H, I are collinear.

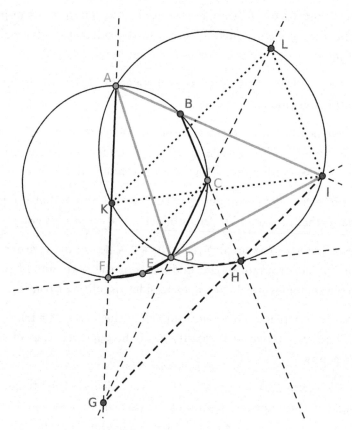

Figure 6.2.45: Pascal's Theorem

Hint: construct $\overline{AD}$. Construct a circle about the triangle $\triangle ADI$, cutting the extended segments $\overline{AF}$, $\overline{CD}$, if necessary, at K and L. Construct $\overline{IK}$, $\overline{KL}$, $\overline{LI}$. By [3.21], we find that $\angle KLG = \angle FCG = \angle GAD$. Therefore $KL \parallel CF$.

Similarly, $\overline{LI} \parallel \overline{CH}$ and $\overline{KI} \parallel \overline{FH}$; hence the triangles $\triangle KLI$, $\triangle FCH$ are homothetic, and so the lines joining corresponding vertices are concurrent. Therefore, the points I, H, G are collinear.

51. If two sides of a triangle are given in position, and if the area is given in magnitude, prove that two points can be found at each of which the base stands opposite a constant angle.

52. If a, b, c, d denote the sides of a cyclic quadrilateral and s its semi-perimeter, prove that its area $= \sqrt{(s-a)(s-b)(s-c)(s-d)}$.

53. If three concurrent lines from the angles of a triangle $\triangle ABC$ meet the opposite side in the points A', B', C', and the points A', B', C' are joined and form a second triangle $\triangle A'B'C'$, prove that

$$\triangle ABC : \triangle A'B'C' = \overline{AB} \cdot \overline{BC} \cdot \overline{CA} : 2 \cdot \overline{AB'} \cdot \overline{BC'} \cdot \overline{CA'}$$

54. In the same case as #53, find the diameter of the circle circumscribed about the triangle $\triangle ABC = \overline{AB'} \cdot \overline{BC'} \cdot \overline{CA'}$ divided by the area of $A'B'C'$.

55. If a quadrilateral is inscribed in one circle and circumscribed to another, the square of its area is equal to the product of its four sides.

56. If on the sides $\overline{AB}$ and $\overline{AC}$ of a triangle $\triangle ABC$ we take two points D and E on their connecting segment such that

$$\frac{BD}{AD} = \frac{AE}{CE} = \frac{DE}{EF}$$

then prove that $\triangle BFC = 2 \cdot \triangle ADE$.

57. If through the midpoints of each of the two diagonals of a quadrilateral we construct a parallel to the other, prove that the lines constructed from their points of intersection to the midpoints of the sides divide the quadrilateral into four equal parts.

58. Suppose that $\overline{CE}$, $\overline{DF}$ are perpendiculars to the diameter of a semicircle, and two circles are constructed touching $\overline{CE}$, $\overline{DE}$, and the semicircle, one internally and the other externally. Prove that the area of the rectangle contained by the perpendiculars from their centers on $\overline{AB}$ is equal to the area $\overline{CE} \cdot \overline{DF}$.

59. If segments are constructed from any point in the circumference of a circle to the vertices of any inscribed regular polygon of an odd number of sides, prove that the sums of the alternate lines are equal.

60. If at the endpoints of a chord constructed through a given point within a given circle tangents are constructed, prove that the sum of the reciprocals of the perpendiculars from the point upon the tangents is constant.

61. If the vertical angle and the bisector of the vertical angle is given, prove that the sum of the reciprocals of the containing sides is constant.

62. If P, P' denote the areas of two regular polygons of any common number of sides inscribed and circumscribed to a circle, and Π, Π' are the areas of the corresponding polygons of double the number of sides, prove that Π is a geometric mean between P and P' and Π' a harmonic mean between Π and P.

63. Prove that the difference of the areas of the triangles formed by joining the centers of the circles constructed about the equilateral triangles constructed outwards on the sides of any triangle is equal to the area of that triangle. Prove the same if they are constructed inwards.

64. In the same case as #63, prove that the sum of the squares of the sides of the two new triangles is equal to the sum of the squares of the sides of the original triangle.

65. Suppose that R and r denote the radii of the circumscribed and inscribed circles to a regular polygon of any number of sides, R', r', corresponding radii to a regular polygon of the same area, and double the number of sides. Prove that $R' = \sqrt{Rr}$ and $r' = \sqrt{\frac{r(R+r)}{2}}$.

66. If the altitude of a triangle is equal to its base, prove that the sum of the distances of the orthocenter from the base and from the midpoint of the base is equal to half the base.

67. Given the area of a parallelogram, one of its angles, and the difference between its diagonals, construct the parallelogram.

68. Given the base of a triangle, the vertical angle, and the point in the base whose distance from the vertex is equal half the sum of the sides, construct the triangle.

69. If the midpoint of the base BC of an isosceles triangle $\triangle ABC$ is the center of a circle touching the equal sides, prove that any variable tangent to the circle will cut the sides in points D, E, such that the rectangle $\overline{BD} \cdot \overline{CE}$ is constant.

70. Inscribe in a given circle a trapezoid, the sum of whose opposite parallel sides is given and whose area is given.

71. Inscribe in a given circle a polygon all of whose sides pass through given points.

72. If two circles $\circ ABC$, $\circ XYZ$ are related such that a triangle may be inscribed in $\circ ABC$ and circumscribed about $\circ XYZ$, prove that an infinite number of such triangles can be constructed.

73. In the same case as #72: prove that the circle inscribed in the triangle formed by joining the points of contact on $\circ XYZ$ touches a given circle.

6.2. PROPOSITIONS FROM BOOK VI

74. In the same case as #72: prove that the circle constructed about the triangle formed by drawing tangents to $\bigcirc ABC$ at the vertices of the inscribed triangle touches a given circle.

75. Find a point, the sum of whose distances from three given points is a minimum.

76. Prove that a line constructed through the intersection of two tangents to a circle is divided harmonically by the circle and the chord of contact.

77. Construct a quadrilateral similar to a given quadrilateral whose four sides pass through four given points.

78. Construct a quadrilateral similar to a given quadrilateral whose four vertices lie on four given lines.

79. Given the base of a triangle, the difference of the base angles, and the rectangle of the sides, construct the triangle.

80. Suppose that $\square ABCD$ is a square, the side CD is bisected at E, and $\overline{EF}$ is constructed making the angle $\angle AEF = \angle EAB$. Prove that $\overline{EF}$ divides the side BC in the ratio of $2:1$.

81. If two circles touch and through their point of intersection two secants be constructed at right angles to each other, cutting the circles respectively in the points A, A'; B, B'; then $(\overline{AA'})^2 + (\overline{BB'})^2$ is constant.

82. If two secants stand at right angles to each other which pass through one of the points of intersection of two circles also cut the circles again, and the line through their centers is the two systems of points $a, b, c; a', b', c'$ respectively, prove that $ab : bc = a'b' : b'c'$.

83. The rectangle contained by the segments of the base of a triangle made by the point of intersection of the inscribed circle is equal to the rectangle contained by the perpendiculars from the endpoints of the base on the bisector of the vertical angle.

84. If O is the center of the inscribed circle of the triangle, prove

$$\frac{OA^2}{bc} + \frac{OB^2}{ca} + \frac{OC^2}{ab} = 1$$

85. State and prove the corresponding theorems for the centers of the escribed circles.

86. Suppose that four points A, B, C, D are collinear. Find a point P at which the segments $\overline{AB}, \overline{BC}, \overline{CD}$ stand opposite equal angles.

87. Prove that the product of the bisectors of the three angles of a triangle whose sides are a, b, c, is

$$\frac{8abc \cdot s \cdot \text{area}}{(a+b)(b+c)(c+a)}$$

88. In the same case as #87, prove that the product of the alternate segments of the sides made by the bisectors of the angles is

$$\frac{a^2b^2c^2}{(a+b)(b+c)(a+c)}$$

89. If three of the six points in which a circle meets the sides of any triangle are such that the lines joining them to the opposite vertices are concurrent, prove that the same property is true of the three remaining points.

90. If a triangle $\triangle A'B'C'$ is inscribed in $\triangle ABC$, prove

$$\overline{AB'} \cdot \overline{BC'} \cdot \overline{CA'} + \overline{A'B} \cdot \overline{B'C} \cdot \overline{C'A}$$

equals twice the area of $\triangle A'B'C'$ multiplied by the diameter of the circle $\bigcirc ABC$.

91. Prove that the medians of a triangle divide each other in the ratio of $2:1$.

Chapter 14

Solutions

14.1 Solutions for Chapter 1

[1.1] Exercises

1. If the segments $\overline{AF}$ and $\overline{BF}$ are constructed, prove that the figure $\square ACBF$ is a rhombus.

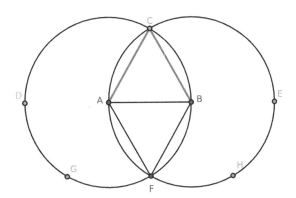

Figure 14.1.1: [1.1, #1]

Proof. Construct $\overline{AF}$ and $\overline{BF}$. By an argument similar to the proof of [1.1], $\overline{AB} = \overline{AF} = \overline{BF}$. Since $\overline{AC} = \overline{AB} = \overline{BC}$ from [1.1],

$$\overline{AC} = \overline{BC} = \overline{BF} = \overline{AF}$$

and so $ACBF$ is equilateral.

By [1.8], $\triangle ABC \cong \triangle ABF$ and since each are equilateral, $\angle CAB = \angle FBA$. By [Cor. 1.29.1], $\overline{AC} \parallel \overline{BF}$. Similarly, $\overline{AF} \parallel \overline{BC}$, and so $ACBF$ is a parallelogram. Since it is also equilateral, by [Def 1.29] $\square ACBF$ is a rhombus. □

[1.1] Exercises

2. If $\overline{CF}$ is constructed and $\overline{AB}$ is extended to the circumferences of the circles (at points D and E), prove that the triangles $\triangle CDF$ and $\triangle CEF$ are equilateral.

Proof. Construct $\overline{CF}$, and extend $\overline{AB}$ to $\overleftrightarrow{AB}$ where $\overleftrightarrow{AB}$ intersects $\odot A$ at D and $\odot B$ at E. Finally, construct $\triangle CDF$ and $\triangle CEF$. We wish to show that $\triangle CDF$ and $\triangle CEF$ are equilateral.

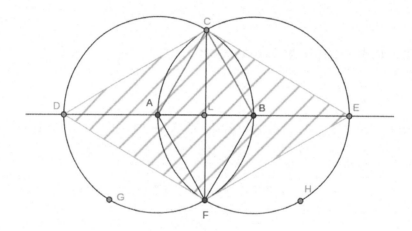

Figure 14.1.2: [1.1, #2]

By [1.1], $\triangle ABC$ is equilateral. By the proof of [1.1, #1][1], $\triangle AFB$ is also equilateral. Since $\overline{AC} = \overline{AF}$, by [1.8] $\triangle ABC \cong \triangle AFB$. Since $\overline{AB}$ is a radius of $\odot A$ and $\odot B$, it follows that the radii of $\odot A$ are equal in length to radii of $\odot B$, and so

$$\overline{AB} = \overline{AC} = \overline{AD} = \overline{AF} = \overline{BC} = \overline{BE} = \overline{BF}$$

Since $\triangle ABC$ is equilateral, $\angle BAC = \angle ABC$. Since $\overline{DE}$ is a segment, $\angle BAC + \angle CAD =$ two right angles $= \angle ABC + \angle CBE$. It follows that $\angle CAD = \angle CBE$. Similarly, we can show that $\angle FAD = \angle FBE$.

---

[1] Notice that it is permissible and encouraged to cite the results of previous exercises. This is the opposite of most K-12 math courses, where students can be punished for treating problems as parts of an interconnected whole. Students should unlearn this "lesson" as soon as possible; math and the sciences are not about isolated pieces, but connections.

What is **not** permissible is circular reasoning: to cite the result of problem #1 in the proof of problem #2 and also cite the result of problem #2 in the proof of problem #1.

That said, some problems in this chapter will be solved without referring to previous solutions if only to prevent this document from becoming more frustrating than it already is.

14.1. SOLUTIONS FOR CHAPTER 1

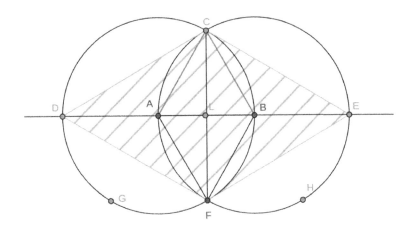

Figure 14.1.3: [1.1, #2]

Since $\triangle ABC \cong \triangle AFB$, $\angle BAC = \angle BAF$, and so $\angle DAF = \angle CAD$. That is,

$$\angle FAD = \angle FBE = \angle DAC = \angle EBC$$

Consider $\triangle ADC$, $\triangle ADF$: $\overline{AF} = \overline{AD} = \overline{AC}$ by the above and $\angle FAD = \angle DAC$. By [1.4], $\triangle ADC \cong \triangle ADF$.

Consider $\triangle ADC$, $\triangle BEC$: $\overline{AC} = \overline{AD} = \overline{BC} = \overline{BE}$ by the above and $\angle CAD = \angle CBE$. By [1.4], $\triangle ADC \cong \triangle BEC$.

Similarly, we can show that $\triangle BEC \cong \triangle BEF$, and so

$$\triangle ADC \cong \triangle ADF \cong \triangle BEF \cong \triangle BEC$$

Hence, $\overline{DF} = \overline{DC} = \overline{CE} = \overline{EF}$.

By [1.32], the sum of the three interior angles of a triangle equals two right angles. To make our calculation easier, define the measure of two right angles to equal π radians.

Let α = each interior angle of $\triangle ABC$. By [1.5.1], $\triangle ABC$ is equiangular, and so $3\alpha = \pi$, or $\alpha = \pi/3$.

By [1.13], $\angle CAL + \angle CAD = \pi$. Since $\angle CAL = \pi/3$, $\angle CAD = 2\pi/3$. Similarly, $\angle FAL = \pi/3$.

Consider $\triangle CAF$, $\triangle CAD$: $\overline{AD} = \overline{AC} = \overline{AF}$ since each are radii of $\odot A$, $\angle CAD = 2\pi/3$, and $\angle CAF = \angle CAL + \angle FAL = 2\pi/3$. By [1.4], $\triangle CAF \cong \triangle CAD$, and so $\overline{CD} = \overline{CF}$.

Hence, $\overline{CD} = \overline{CF} = \overline{DF}$, or $\triangle CDF$ is equilateral. By the above and [1.8], $\triangle CDF \cong \triangle CEF$, which completes the proof. □

Corollary. *[1.1, #1.1]* □*CDFE IS A RHOMBUS.*

Applying [1.1, #1] to □*CDFE, we find that* □*CDFE is a rhombus.*

[1.2] Exercises

1. Prove [1.2] when A is a point on $\overline{BC}$.

Given a point on an arbitrary segment, it is possible to construct a segment with:

  (1) one endpoint being the previously given point

  (2) its length is equal to that of the arbitrary segment.

Proof. Let $\overline{BC}$ be an arbitrary segment where A is a point on $\overline{BC}$. We claim that we can construct a segment with A as an endpoint such that its length is equal to that of $\overline{BC}$.

If $A = B$ or $A = C$, the proof is trivial.[2]

Suppose that A is not an endpoint of $\overline{BC}$. Construct the equilateral triangle $\triangle ABD$ [1.1]. Also construct the circle $\odot A$ with radius equal in length to $\overline{AC}$. Extend side DA to $\overrightarrow{DA}$ where $\overrightarrow{DA}$ intersects $\odot A$ at E.

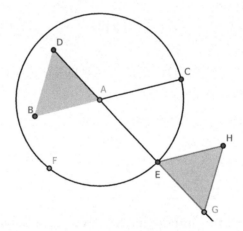

Figure 14.1.4: [1.2, #1]

Construct the equilateral triangle $\triangle EGH$ where G is also a point on $\overrightarrow{DA}$ and $\overline{EG} = \overline{AD}$ [1.2]. So $\overline{AE} \oplus \overline{EG} = \overline{AG}$ and

$$\begin{aligned} \overline{AG} &= \overline{AE} + \overline{EG} \\ &= \overline{AC} + \overline{AD} \\ &= \overline{AC} + \overline{AB} \\ &= \overline{BC} \end{aligned}$$

---

[2]A trivial proof is one that is immediately obvious. In this case, if $A = B$ or $A = C$, then $\overline{BC}$ itself is the segment we require. Since it already exists, there is nothing to do but acknowledge that the proof took no effort on our part, i.e., it is trivial.

14.1. SOLUTIONS FOR CHAPTER 1

which completes the proof. □

[1.4] Exercises

Prove the following:

1. The line that bisects the vertical angle of an isosceles triangle also bisects the base perpendicularly.

Proof. Suppose $\triangle ABC$ is an isosceles triangle where $AB = AC$. Further suppose that the ray AD bisects $\angle BAC$.[3] We claim that $\overline{BD} = \overline{CD}$ and $\overline{AD} \perp \overline{BC}$.

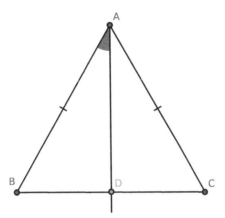

Figure 14.1.5: [1.4, #1]

Consider $\triangle ABD$, $\triangle ACD$: the triangles share side AD, and by hypothesis $AB = AC$ and $\angle DAB = \angle DAC$. By [1.4], $\triangle ABD \cong \triangle ACD$. Hence, $\overline{BD} = \overline{CD}$. Also, $\angle ADB = \angle ADC$; since they are supplements, they stand at right angles [Def. 1.14], and so, $\overline{AD} \perp \overline{BC}$. □

2. If two adjacent sides of a quadrilateral are equal in length and the diagonal bisects the angle between them, then their remaining sides are also equal in length.

Proof. Suppose that $ABCD$ is a quadrilateral where $AB = AC$ and where the diagonal AD bisects $\angle BAC$. We claim that $BD = CD$.

Consider $\triangle ACD$, $\triangle ABD$: since $AC = AB$, each shares side AD, and $\angle DAC = \angle DAB$ by hypothesis, by [1.4] $\triangle ACD \cong \triangle ABD$. Therefore, $BD = CD$.

---

[3]A line or a segment of appropriate size may be substituted, *mutatis mutandis*.

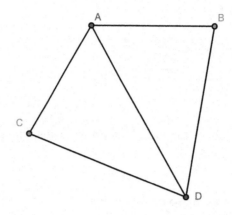

Figure 14.1.6: [1.4, #2]

3. If two segments stand perpendicularly and each bisects the other, then any point on one segment is equally distant from the endpoints of the other segment.

Proof. Suppose $\overline{AB} \perp \overline{CD}$ and that $\overline{AB}$ and $\overline{CD}$ bisect each other at E. Wlog, let F be a point on AB. We claim that F is equidistant from C and D.

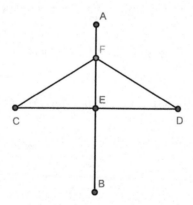

Figure 14.1.7: [1.4, #3]

Construct $\triangle CEF$ and $\triangle DEF$. Since $\overline{AB} \perp \overline{CD}$ and $\overline{AB}$ and $\overline{CD}$ bisect each other at E by hypothesis, $\angle CEF = \angle DEF$ and $\overline{CE} = \overline{DE}$. Since $\triangle CEF$ and $\triangle DEF$ share side FE, by [1.4] we find that $\triangle CEF \cong \triangle DEF$ Hence, $CF = DF$.

The proof for any point on CD is similar to the above, *mutatis mutandis*. Therefore, we have proven our claim. □

14.1. SOLUTIONS FOR CHAPTER 1

Corollary. *to [1.4, #3]:* $\overleftrightarrow{AB}$ *is the Axis of Symmetry of* $\triangle CFD$.

[1.5] Exercises

2. Prove that $\overleftrightarrow{AH}$ is an Axis of Symmetry of $\triangle ABC$.

Proof. Construct the figure from [1.5], construct $\overleftrightarrow{AH}$, and let I be the intersection of $\overline{BC}$ and $\overleftrightarrow{AH}$. We claim that $\overleftrightarrow{AH}$ is the Axis of Symmetry of $\triangle ABC$.

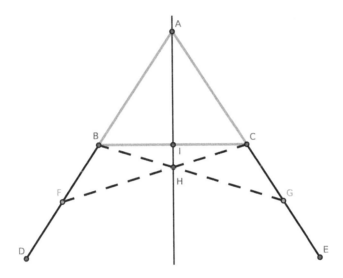

Figure 14.1.8: [1.5, #2]

Consider $\triangle BHF$ and $\triangle CHG$: by the proof of [1.5] $\angle BFH = \angle CGH$ and $BF = CG$. By [1.15] $\angle BHF = \angle CHG$, and so by [1.26], $\triangle BHF \cong \triangle CHG$. It follows that $BH = CH$.

Consider $\triangle ABH$ and $\triangle ACH$: by the above, $BH = CH$; by the proof of [1.5] $AB = AC$; finally, the triangles share side AH. By [1.8], $\triangle ABH \cong \triangle ACH$. It follows that $\angle BAH = \angle CAH$.

Consider $\triangle ABI$ and $\triangle ACI$: by the above $\angle BAI = \angle CAI$, $AB = AC$, and the triangles share side AI. By [1.4], $\triangle ABI \cong \triangle ACI$. It follows that $\triangle ABI = \triangle ACI$.

Since $\triangle ABC = \triangle ABI \oplus \triangle ACI$, we find that $\overleftrightarrow{AH}$ is the Axis of Symmetry of $\triangle ABC$ by [Def 1.35]. This proves our claim. □

[1.5] Exercises

4. Take the midpoint on each side of an equilateral triangle; the segments joining them form a second equilateral triangle.

Proof. Suppose that $\triangle ABC$ is equilateral. Construct the midpoints I, J, and K on sides BC, AB, and AC, respectively. We claim that $\triangle IJK$ is equilateral.

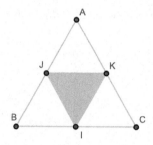

Figure 14.1.9: [1.5, #4]

Since I is the midpoint of side BC, $\overline{IB} = \overline{IC}$. Similarly, $\overline{JA} = \overline{JB}$. Since $\triangle ABC$ is equilateral, $\overline{IB} = \overline{JB}$. Continuing in this way, we can show that

$$\overline{IB} = \overline{JB} = \overline{JA} = \overline{AK} = \overline{KC} = \overline{IC}$$

And by [1.5, Cor. 1] we also have that

$$\angle ABC = \angle ACB = \angle BAC$$

Hence by multiple applications of [1.4],

$$\triangle JBI \cong \triangle KCI \cong \triangle JAK$$

It follows that $\overline{IJ} = \overline{JK} = \overline{KI}$, and so $\triangle IJK$ is equilateral. This proves our claim. $\square$

14.1. SOLUTIONS FOR CHAPTER 1

[1.9] Exercises

2. Prove that $\overline{AF} \perp \overline{DE}$.

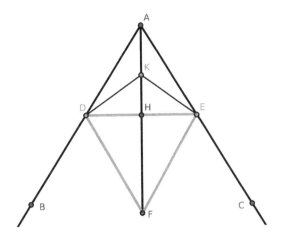

Figure 14.1.10: [1.9, #2]

Proof. Consider $\triangle ADH$, $\triangle AEH$: by construction in [1.9], $\angle DAH = \angle EAH$ and $\overline{AD} = \overline{AE}$, and each triangle shares side AH. By [1.4], $\triangle ADH \cong \triangle AEH$. It follows that $\overline{DH} = \overline{HE}$. Also, $\angle DHA = \angle EHA$. Since these are adjacent angles, they are right angles [Def. 1.14]. By [1.15], $\angle DHF$ and $\angle EHF$ are also right angles. Thus, $\overline{AF} \perp \overline{DE}$, which concludes the proof. $\square$

3. Prove that any point on $\overline{AF}$ is equally distant from points D and E.

Proof. Construct K on $\overline{AF}$. We claim that $\overline{DK} = \overline{KE}$.

Consider $\triangle DHK$ and $\triangle EHK$: by [1.9, #2], $\overline{DH} = \overline{HE}$, the triangles share $\overline{HK}$, and $\angle DHK = \angle EHK$. By [1.4], $\triangle DHK \cong \triangle EHK$. It follows that $\overline{DK} = \overline{KE}$, and which proves our claim. $\square$

[1.10] Exercises

1. Bisect a segment by constructing two circles.

Proof. Construct the figure from [1.1, #2]. We claim that $\overline{CF}$ bisects $\overline{AB}$.

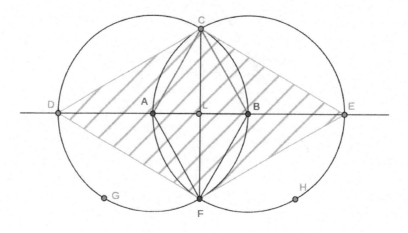

Figure 14.1.11: [1.1, #2]

Since $\square DCEF = \triangle CDF \oplus \triangle CEF$ and $\triangle CDF \cong \triangle CEF$, we find that $\overline{CF}$ is an axis of symmetry of $\square DCEF$ [Def. 1.35]. It follows that $\overline{DL} = \overline{LE}$. Since $\overline{DA} = \overline{BE}$ by the proof of [1.1, #2], $\overline{AL} = \overline{LB}$, which completes the proof. □

14.1. SOLUTIONS FOR CHAPTER 1

[1.10] Exercises

2. Extend $\overline{CD}$ to $\overleftrightarrow{CD}$. Prove that every point equally distant from the points A and B are points on $\overleftrightarrow{CD}$.

Proof. Extend $\overline{CD}$ to $\overleftrightarrow{CD}$ and construct E on $\overleftrightarrow{CD}$. We claim that E is equally distant from A and B.

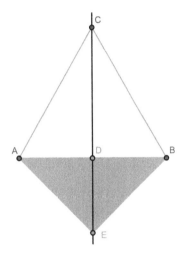

Figure 14.1.12: [1.10, #2]

By the proof of [1.10], $\triangle ACB$ is isosceles. By [Cor. 1.9.1], $\overleftrightarrow{CD}$ is an Axis of Symmetry to $\triangle ACB$. It follows that $\overleftrightarrow{CD}$ is an Axis of Symmetry to any triangle with vertices A, B, and E, where E is any point on $\overleftrightarrow{CD}$ except D itself. Hence, $\overline{AE} = \overline{EB}$ for any point E on $\overleftrightarrow{CD}$. Since $\overline{AD} = \overline{DB}$ by [1.10], we have proven that every point on $\overleftrightarrow{CD}$ is equally distant from A and B.

Suppose point K exists where $\overline{AK} = \overline{BK}$ but K is not on $\overleftrightarrow{CD}$. It follows that $\triangle ABE$ is distinct from $\triangle ABK$ where the triangles share side AB, $AE = AK$, and $BE = BK$. But this construction contradicts [1.7]. Hence, no such K exists, which completes the proof. □

[1.11] Exercises

1. Prove that the diagonals of a rhombus bisect each other perpendicularly.

Proof. Construct the figure from [1.11], and also construct the equilateral triangle $\triangle DEG$ where G lies on the opposite side of $\overline{AB}$ from F. By the proof of [1.11], $\triangle DFE$ is equilateral. By [1.8], $\triangle DFE \cong \triangle DGE$.

Construct $\overline{GC}$; by the proof of [1.11], $GC \perp AB$. We claim that $\square FEGD$ is a rhombus, $\overline{GF}$ and $\overline{DE}$ are its diagonals, and $\overline{GF}$ and $\overline{DE}$ bisect each other.

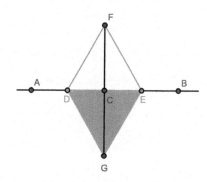

Figure 14.1.13: [1.11, #1]

Consider $\triangle DCF$ and $\triangle DCG$: since $\triangle DEF \cong \triangle DEG$, $DF = DG$ and $\angle CDG = \angle CDF$; the triangles also share side DC. By [1.4], $\triangle DCF \cong \triangle DCG$. Similarly, we can show that $\triangle ECF \cong \triangle ECG$.

By the proof to [1.11], $\triangle DCF \cong \triangle ECF$. By the above, it can be shown that

$$\triangle DCF \cong \triangle DCG \cong \triangle ECF \cong \triangle ECG$$

Hence, $\overline{FD} = \overline{FE} = \overline{GD} = \overline{GE}$, and so $FEGD$ is a equilateral.

Since $\angle DGF = \angle EFG$ by the above, $\overline{DG} \parallel \overline{EF}$; since $\angle EGF = \angle DFG$, $\overline{EG} \parallel \overline{FD}$. Hence, $\square FEGD$ is an equilateral parallelogram. By [Def. 1.29], $\square FEGD$ is a rhombus. Clearly, $\overline{DE}$ and $\overline{GF}$ are the diagonals of $\square FEGD$.

By the above, $\overline{DC} = \overline{CE}$ and $\overline{FC} = \overline{CG}$, so the diagonals of $\square FEGD$ bisect each other. By [1.11], $\angle DCF$ is a right angle, and so $\overline{DE} \perp \overline{GF}$, which completes the proof. $\square$

14.1. SOLUTIONS FOR CHAPTER 1

[1.11] Exercises

3. Find a point on a given line that is equally distant from two given points.

Proof. Let $\overleftrightarrow{AB}$ be a given line, and let C, D be points not on $\overleftrightarrow{AB}$. We wish to find a point F on $\overleftrightarrow{AB}$ such that $\overline{FC} = \overline{FD}$.

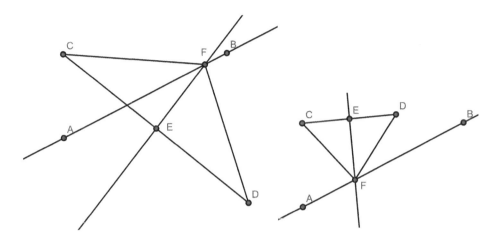

Figure 14.1.14: [1.11, #3]

Construct $\overline{CD}$; by [1.10], locate its midpoint, E. Construct $\overleftrightarrow{FE}$ such that $\overline{CD} \perp \overleftrightarrow{FE}$ and F is a point on $\overleftrightarrow{AB}$. We claim that F is equally distant from C and D.

Consider $\triangle CEF$ and $\triangle DEF$: $CE = DE$ by construction, $\angle CEF = \angle DEF$ by construction, and the triangles share side EF. By [1.4], $\triangle CEF \cong \triangle DEF$. Hence, $\overline{CF} = \overline{DF}$, which completes the proof. □

[1.12] Exercises

1. Prove that circle $\odot C$ cannot intersect $\overleftrightarrow{AB}$ at more than two points.

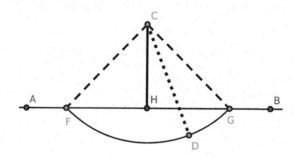

Figure 14.1.15: [1.12, #1]

Proof. Suppose $\odot C$ intersects $\overleftrightarrow{AB}$ at more than two points. If the third point lies between points F and G, then the radius of $\odot C$ must decrease in length; this contradicts the definition of a radius (that it must have a fixed length).

Similarly, if the third point lies to the left of F or to the right of G, the radius of $\odot C$ must increase in length, resulting in a similar contradiction.

Hence, $\odot C$ cannot intersect $\overleftrightarrow{AB}$ at more than two points. $\square$

[1.19] Exercises

3. Prove that three equal and distinct segments cannot be constructed from the same point to the same line.

Proof. Suppose such a construction were possible. Consider the common point to be the center of a circle and the three equal yet distinct segments to be radii of that circle. Then we could construct a line such that the circumference of the circle intersects the line at three points. This contradicts [1.12, #1], which completes the proof. $\square$

14.1. SOLUTIONS FOR CHAPTER 1

[1.19] Exercises

5. If $\triangle ABC$ is a triangle such that $AB \leq AC$, then a segment $\overline{AG}$, constructed from A to any point G on side BC, is less than AC.

Proof. Construct $\triangle ABC$ where $AB < AC$. Construct $\overline{AG}$ where G is a point on side BC (other than B and C). We claim that $\overline{AG} < \overline{AC}$.

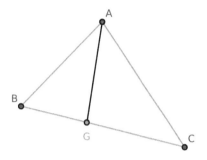

Figure 14.1.16: [1.19, #5]

If $\overline{AB} = \overline{AC}$, then by [1.19, #3], $\overline{AG} < \overline{AC}$.

If $\overline{AB} < \overline{AC}$, extend $\overline{AB}$ to $\overline{AH}$ such that $\overline{AH} = \overline{AC}$ and construct $\overline{CH}$. Extend $\overline{AG}$ to $\overline{AJ}$ where J is on $\overline{CH}$. Clearly, $\overline{AG} \leq \overline{AJ}$.

Since $\overline{AH} = \overline{AC}$, by [1.19, #3], $\overline{AJ} < \overline{AC}$. Since $\overline{AG} \leq \overline{AJ}$, $\overline{AG} < \overline{AC}$, which completes the proof. □

[1.20] Exercises

5. The perimeter of a quadrilateral is greater than the sum of its diagonals.

Proof. Suppose that $ABCD$ is a quadrilateral with diagonals $\overline{AC}$ and $\overline{BD}$. We claim that
$$\overline{AB} + \overline{BC} + \overline{CD} + \overline{DA} > \overline{AC} + \overline{BD}$$

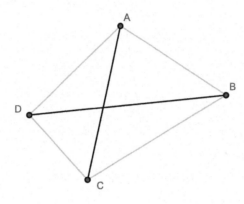

Figure 14.1.17: [1.20, #5]

By [1.20], we have
$$\begin{aligned} AD + CD &> AC \\ AB + BC &> AC \\ AD + AB &> BD \\ BC + CD &> BD \end{aligned}$$

Or,
$$2 \cdot (AB + BC + CD + DA) > 2 \cdot (AC + BD)$$
$$\Longrightarrow$$
$$AB + BC + CD + DA > AC + BD$$

which proves our claim. $\square$

[1.20] Exercises

6. The sum of the lengths of the three medians of a triangle is less than $\frac{3}{2}$ times its perimeter.

Proof. Construct $\triangle ABC$ with medians $\overline{AF}$, $\overline{BE}$, and $\overline{CD}$. We claim that

$$\overline{AF} + \overline{BE} + \overline{CD} < \frac{3}{2} \cdot (\overline{AB} + \overline{BC} + \overline{AC})$$

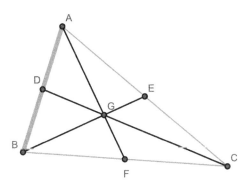

Figure 14.1.18: [1.20, #6]

Consider $\triangle ABE$: by [1.20], we have

$$\overline{AE} + \overline{AB} > \overline{BE}$$

Similarly, in $\triangle DBC$, we have

$$\overline{BD} + \overline{BC} > \overline{CD}$$

and in $\triangle ACF$, we have

$$\overline{AC} + \overline{CF} > \overline{AF}$$

Recall that $\overline{AE} = \frac{1}{2}\overline{AC}$, $\overline{BD} = \frac{1}{2}\overline{AB}$, and $\overline{CF} = \frac{1}{2}\overline{BC}$. Adding each inequality, we find that

$$\overline{AB} + \overline{BC} + \overline{AC} + \overline{AE} + \overline{BD} + \overline{CF} > \overline{AF} + \overline{BE} + \overline{CD}$$

$$\overline{AB} + \overline{BC} + \overline{AC} + \tfrac{1}{2}\overline{AB} + \tfrac{1}{2}\overline{BC} + \tfrac{1}{2}\overline{AC} > AF + BE + CD$$

$$\tfrac{3}{2}(\overline{AB} + \overline{BC} + \overline{AC}) > \overline{AF} + \overline{BE} + \overline{CD}$$

which proves our claim. □

[1.23] Exercises

1. Construct a triangle given two sides and the angle between them.

Proof. Suppose we have arbitrary segments $\overline{AB}$ and $\overline{CD}$ and an arbitrary angle $\angle EFG$. We claim that we can construct $\triangle HMN$ from $\overline{AB}, \overline{CD}$, and $\angle EFG$.

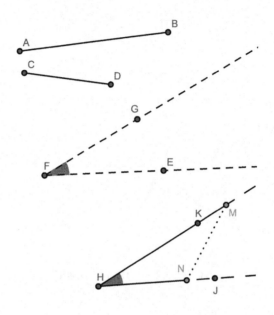

Figure 14.1.19: [1.23, #1]

By [1.23], construct rays $\overrightarrow{HJ}$ and $\overrightarrow{HK}$ such that $\angle JHK = \angle EFG$. Construct $\overline{HM}$ on $\overrightarrow{HJ}$ and $\overline{HN}$ on $\overrightarrow{HK}$ such that $\overline{AB} = \overline{HM}$ and $\overline{CD} = \overline{HN}$. Construct $\overline{MN}$. Notice that $\triangle MNH$ has sides equal in length to segments $\overline{AB}$ and $\overline{CD}$ and contains an angle equal in measure to $\angle EFG$. This proves our claim. $\square$

[1.29] Exercises

2. Construct $\overleftrightarrow{AB}$ containing the point C and $\overleftrightarrow{EF}$ containing the point D such that $\overleftrightarrow{AB} \parallel \overleftrightarrow{EF}$. Construct $\overline{CH}$ and $\overline{CJ}$ such that $\overline{CJ}$ bisects $\angle ACD$ and $\overline{CH}$ bisects $\angle BCD$. Prove that $\overline{DH} = \overline{DJ}$.

Proof. Our hypothesis and claim are stated above. Construct the above as well as $\overleftrightarrow{JK}$ and $\overleftrightarrow{HL}$ such that $\overleftrightarrow{JK} \parallel \overline{CD}$ and $\overline{CD} \parallel \overleftrightarrow{HL}$. By [1.30], $\overleftrightarrow{JK} \parallel \overleftrightarrow{HL}$.

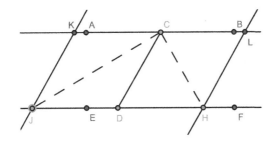

Figure 14.1.20: [1.29, #2]

Since $\overline{CH}$ is a bisector of $\angle BCD$, $\angle BCH = \angle DCH$. Since $\overline{CD} \parallel \overleftrightarrow{HL}$, $\angle DCH = \angle CHL$. And since $\overleftrightarrow{AB} \parallel \overleftrightarrow{EF}$, $\angle BCH = \angle DHC$. This gives us

$$\angle BCH = \angle DCH = \angle DHC = \angle CHL$$

Consider $\triangle CDH$. Since $\angle DCH = \angle DHC$, by [1.6] $\overline{DC} = \overline{DH}$.

Similarly, it can be shown that

$$\angle JCD = \angle JCK = \angle DJC = \angle DCJ$$

and so $\overline{DJ} = \overline{DC}$.

Thus, $\overline{DJ} = \overline{DH}$, which completes the proof. $\square$

[1.29] Exercises

5. Two lines passing through a point which is equidistant from two parallel lines intercept equal segments on the parallels.

Proof. Construct $\overleftrightarrow{AB}$ and $\overleftrightarrow{CD}$ such that $\overleftrightarrow{AB} \parallel \overleftrightarrow{CD}$. Construct $\overleftrightarrow{LM}$ such that $\overleftrightarrow{AB} \perp \overleftrightarrow{LM}$ and bisect $\overleftrightarrow{LM}$ at G [1.10]. Construct arbitrary lines $\overleftrightarrow{HJ}$ and $\overleftrightarrow{IK}$ such that each passes through G. We claim that $\overline{HI} = \overline{JK}$.

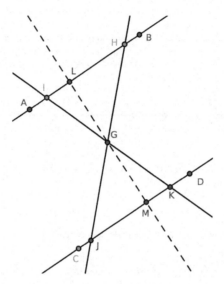

Figure 14.1.21: [1.29, #5]

Consider $\triangle GLI$ and $\triangle GMK$: $\angle LGI = \angle MGK$ by [1.15]; $GL = GM$ by construction; $\angle GMK = \angle GLI$ by construction. By [1.26], $\triangle GLI \cong \triangle GMK$, and so $GI = GK$.

Now consider $\triangle GHI$ and $\triangle GJK$: by the above, $GI = GK$. $\angle HGI = \angle JGK$ by [1.15], and since $AB \parallel CD$, $\angle GIH = \angle GKJ$ by [1.29, Cor. 1]. By [1.26], we find that $\triangle GHI \cong \triangle GJK$. Therefore, $\overline{HI} = \overline{JK}$, which proves our claim. □

14.1. SOLUTIONS FOR CHAPTER 1

[1.31] Exercises

1. Given the altitude of a triangle and the base angles, construct the triangle.

Proof. We propose to construct the triangle with altitude h and base angles α and β.

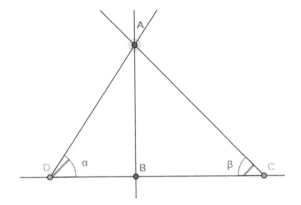

Figure 14.1.22: [1.31, #1]

Let $\overline{AB} = h$. Extend $\overline{AB}$ to $\overleftrightarrow{AB}$, and construct $\overleftrightarrow{BC}$ such that $\overleftrightarrow{AB} \perp \overleftrightarrow{BC}$ [Cor. 1.11.1].

Since $\overleftrightarrow{AB} \perp \overleftrightarrow{BC}$, $\angle DBA = \pi/2$ radians (a right angle). Construct $\overrightarrow{AD}$ such that D is a point on $\overleftrightarrow{BC}$ and $\angle DAB = \frac{\pi}{2} - \alpha$. Then $\angle BDA = \alpha$ [1.32].

Similarly, construct $\overrightarrow{CA}$ so that $\angle BAC = \frac{\pi}{2} - \beta$. Then $\angle BCA = \beta$.

This constructs $\triangle ACD$ with altitude h and interior angles α and β. □

[1.32] Exercises.

3. If the line which bisects an external vertical angle of a triangle is parallel to the base of the triangle, then the triangle is isosceles.

Proof. Construct $\triangle ABC$ with external vertical angle $\angle ACE$ such that $\overleftrightarrow{CD}$ bisects $\angle ACE$ and $\overleftrightarrow{CD} \parallel \overleftrightarrow{AB}$. We claim that $\triangle ABC$ is isosceles.

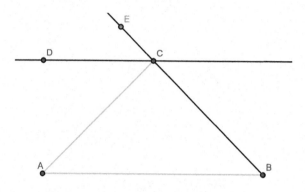

Figure 14.1.23: [1.32, #3]

By [1.29, Cor. 1], $\angle DCE = \angle ABC$. Also by By [1.29, Cor. 1], $\angle DCA = \angle BAC$. But $\angle DCE = \angle DCA$ by hypothesis, and so $\angle ABC = \angle BAC$. By [1.6], $\triangle ABC$ is isosceles. □

[1.32] Exercises

5. Prove that the three altitudes of a triangle are concurrent. [Note: We are proving the existence of the orthocenter of a triangle: the point where the three altitudes intersect, and one of a triangle's points of concurrency.]

Proof. Construct the following: $\triangle ABC$, altitudes $\overline{AG}$ and $\overline{CF}$, and also $\overline{BD}$ where D is the intersection of $\overline{AG}$ and $\overline{CF}$. Extend $\overline{BD}$ to $\overline{BE}$ where E is a point on $\overline{AC}$. We claim that $\overline{BE}$ is an altitude of $\triangle ABC$.

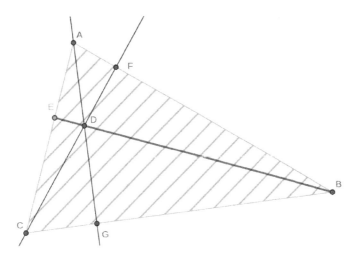

Figure 14.1.24: [1.32, #5]

By the construction of $\angle BDA$, we have $\angle EDB = \angle BDA + \angle EDA$. Consider $\triangle AED$: by [1.32] $\angle BDA = \angle DEA + \angle EAD$. Applying [1.32] again, we obtain

$$\begin{aligned} \angle EDB &= \angle BDA + \angle EDA \\ &= \angle DEA + \angle EAD + \angle EDA \\ &= \pi \text{ radians} \end{aligned}$$

since $\angle DEA$, $\angle EAD$, and $\angle EDA$ are the interior angles of $\triangle EDA$. Hence, $\overline{BE} = \overline{BD} \oplus \overline{DE}$. It remains to be shown that $\angle BEA$ is a right angle.

Suppose $\angle BEA < \angle BEC$. It follows that $\overline{AC}$ is not a straight segment, which contradicts its construction as a side of $\triangle ABC$. We obtain a similar result if $\angle BEA > \angle BEC$. Hence, $\angle BEA = \angle BEC$. Since the angles are adjacent, each are right angles.

Thus, $\overline{EB}$ is an altitude of $\triangle ABC$, and the three altitudes of $\triangle ABC$ are concurrent. This completes the proof. □

[1.32] Exercises

6. The bisectors of the adjacent angles of a parallelogram stand at right angles.

Proof. Construct $\square ABDC$. Also construct $\angle CBD$ such that it bisects $\angle ABD$ as well as $\angle CAD$ such that it bisects $\angle CAB$. We wish to show that $\overline{CB} \perp \overline{AD}$.

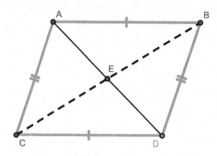

Figure 14.1.25: [1.32], #6

By [Cor. 1.29.1], $\angle DCB = \angle CBA$. By hypothesis, $\angle CBA = \angle CBD$, and so $\angle DCB = \angle CBD$. We may continue this line of reasoning until we obtain

$$\angle CBA = \angle CBD = \angle DCB = \angle ACB$$

and

$$\angle ADC = \angle ADB = \angle DAC = \angle DAB$$

Consider $\triangle CED$, $\triangle CEA$: $\angle ECD = \angle ECA$ (since $\angle DCB = \angle ACB$), $\angle EDC = \angle EAC$ (since $\angle ADC = \angle DAC$), and each shares side CE. By [1.26], $\triangle CED \cong \triangle CEA$. Hence, $\angle CED = \angle CEA$. Since $\angle CED$, $\angle CEA$ are adjacent, they are right angles. It follows that $\overline{CB} \perp \overline{AD}$, which completes the proof. $\square$

[1.33] Exercises

1. Prove that if two segments $\overline{AB}$, $\overline{BC}$ are respectively equal and parallel to two other segments $\overline{DE}$, $\overline{EF}$, then the segment $\overline{AC}$ joining the endpoints of the former pair is equal in length to the segment $\overline{DF}$ joining the endpoints of the latter pair.

Proof. Construct segments $\overline{AB}$, $\overline{DE}$, $\overline{BC}$, and $\overline{EF}$ such that $\overline{AB} = \overline{DE}$, $\overline{BC} = \overline{EF}$, $\overline{AB} \parallel \overline{DE}$, and $\overline{BC} \parallel \overline{EF}$. Construct segments $\overline{AC}$ and $\overline{DF}$. We wish to show $\overline{AC} = \overline{DF}$.

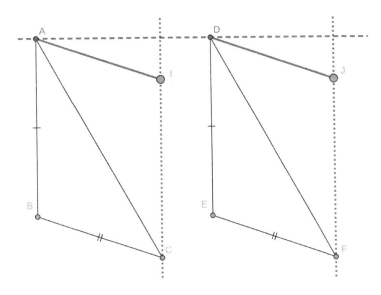

Figure 14.1.26: [1.33, #1]

Construct $\overleftrightarrow{IC}$ such that $\overleftrightarrow{IC} \parallel \overline{AB}$ and $\overleftrightarrow{JF}$ such that $\overleftrightarrow{JF} \parallel \overline{DE}$. Also construct $\overline{AI}$ such that $\overline{AI} \parallel \overline{BC}$ and $\overline{DJ}$ such that $\overline{DJ} \parallel \overline{EF}$.

Suppose $\angle ABC < \angle DEF$. It follows that $\overleftrightarrow{BC}$ and $\overleftrightarrow{EF}$ intersect at a point; this contradicts the construction that $\overline{BC} \parallel \overline{EF}$. A similar contradiction arises if $\angle ABC > \angle DEF$. Hence, $\angle ABC = \angle DEF$.

Consider $\triangle ABC$, $\triangle DEF$: $\overline{AB} = \overline{DE}$, $\angle ABC = \angle DEF$, and $\overline{BC} = \overline{EF}$. By [1.4], $\triangle ABC \cong \triangle DEF$, and so $\overline{AC} = \overline{DF}$. □

[1.34] Exercises

1. Show that the diagonals of a parallelogram bisect each other.

Proof. Consider $\square ABCD$ and diagonals $\overline{AD}, \overline{BC}$. Let point E be the intersection of $\overline{AD}$ and $\overline{BC}$. We claim that $\overline{AE} = \overline{ED}$ and $\overline{CE} = \overline{EB}$.

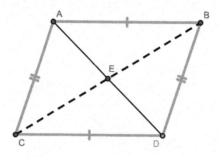

Figure 14.1.27: [1.34], #1

Since $\overline{AB} \parallel \overline{CD}$, $\angle BCD = \angle CBA$. Similarly, we find that $\angle CDA = \angle DAB$. Consider $\triangle ECD$ and $\triangle AEB$: since $\angle ECD = \angle EBA$, $\angle EDC = \angle EAB$, and $\overline{CD} = \overline{AB}$, by [1.26] we find that $\triangle ECD \cong \triangle AEB$. Hence, $\overline{AE} = \overline{ED}$.

A similar argument shows that $\overline{CE} = \overline{EB}$, *mutatis mutandis*, which proves our claim. $\square$

[1.34] Exercises

2. If the diagonals of a parallelogram are equal, each of its angles are right angles.

Proof. Construct $\square ABCD$ and suppose that $\overline{AD} = \overline{BC}$. We wish to show that each interior angle of $\square ABCD$ is a right angle.

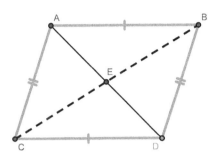

Figure 14.1.28: [1.34], #2

By [1.34], $\overline{AC} = \overline{BD}$ and $\overline{AB} = \overline{CD}$. By [1.34, #1], E bisects both $\overline{AD}$ and $\overline{BC}$. Since $\overline{AD} = \overline{BC}$ by hypothesis,
$$\overline{AE} = \overline{ED} = \overline{CE} = \overline{EB}$$

By [1.32, #6], $\angle CED = \angle DEB = \angle BEA = \angle AEC = \frac{\pi}{2}$ radians in measure. By [1.4],
$$\triangle ECD \cong \triangle EDB \cong \triangle EBA \cong \triangle EAC$$

and so
$$\overline{AB} = \overline{BD} = \overline{DC} = \overline{CA}$$

Consider $\triangle ACD$, $\triangle BAC$: $\overline{AD} = \overline{BC}$ by hypothesis and their sides are equal by the above. By [1.8], $\triangle ACD \cong \triangle BAC$. It follows that $\angle ACD = \angle BAC$. By [1.34], $\angle ACD = \angle ABD$. Clearly,
$$\angle ACD = \angle ABD = \angle BAC = \angle BDC$$

By [1.29], we find that the sum of the interior angles of $\square ABCD = 2\pi$ radians (four right angles). Since each angle equals $\angle ACD$ in measure, $\angle ACD = \pi/2$ radians; or, each interior angle of $\square ABCD$ is a right angle, which completes the proof. $\square$

[1.37] Exercises

1. If two triangles of equal area stand on the same base but on opposite sides of the base, the segment connecting their vertices is bisected by the base or its extension.

Proof. Suppose $\triangle ABG$ and $\triangle ABI$ share base $\overline{AB}$ such that G stands on the opposite side of $\overline{AB}$ than I. Also suppose $\triangle ABG = \triangle ABI$. We claim that $\overline{GI}$ is bisected by $\overleftrightarrow{AB}$.

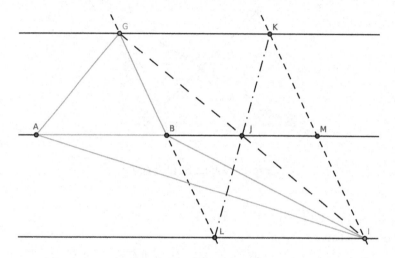

Figure 14.1.29: [1.37, #1]

Extend $\overline{AB}$ to $\overleftrightarrow{AB}$, and let J be the point where $\overline{GI}$ intersects $\overleftrightarrow{AB}$. Construct $\overleftrightarrow{GK}$ and $\overleftrightarrow{LI}$ such that $\overleftrightarrow{GK} \parallel \overleftrightarrow{AB}$ and $\overleftrightarrow{LI} \parallel \overleftrightarrow{AB}$. By [1.30], $\overleftrightarrow{GK} \parallel \overleftrightarrow{LI}$. Also construct $\overleftrightarrow{GB}$ and $\overleftrightarrow{KI}$ such that $\overleftrightarrow{GB} \parallel \overleftrightarrow{KI}$. Hence, $\square GKIL$ is a parallelogram.

Construct $\overline{JL}$ and $\overline{JK}$. If $\angle LJK = \pi$ radians in measure, then LK is a segment. Since $\overline{GI}$ is a segment, $\angle GJL + \angle LJI = \pi$ radians. By [1.15], $\angle KJI = \angle GJL$, and so $\angle LJK = \angle KJI + \angle LJI = \pi$ radians. Hence, $\overline{LK} = \overline{JL} \oplus \overline{JK}$. Furthermore, $\overline{LK}$ and $\overline{GJ}$ are diagonals of $\square GKIL$.

By [1.34, #1], $\overline{GJ} = \overline{JI}$. Since $\overline{GI} = \overline{GJ} \oplus \overline{JI}$, $\overline{GI}$ is bisected by $\overleftrightarrow{AB}$ at J, which completes the proof. □

14.1. SOLUTIONS FOR CHAPTER 1

[1.38] Exercises

1. Every median of a triangle bisects the triangle.

Proof. Construct $\triangle ABC$ where $\overline{AF}$ is the median of side BC. We claim that $\triangle ABF = \triangle ACF$.

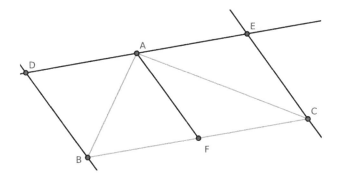

Figure 14.1.30: [1.38, #1]

Construct $\overleftrightarrow{AE}$ such that $\overleftrightarrow{AE} \parallel \overline{BC}$. Clearly, $\triangle ABF$ and $\triangle ACF$ stand between the same parallels. Since $\overline{BF} = \overline{FC}$ by hypothesis, by [1.38] we find that $\triangle ABF = \triangle ACF$. This completes the proof. □

Remark. We do not claim that the triangles are congruent, merely equal in area.

[1.38] Exercises

5. One diagonal of a quadrilateral bisects the other if and only if the diagonal also bisects the quadrilateral.

Proof. Construct quadrilateral $ABCD$ and diagonals $\overline{AC}$ and $\overline{BD}$.

Claim 1: If $\overline{BD}$ bisects $\overline{AC}$, then $\overline{BD}$ bisects $ABCD$.

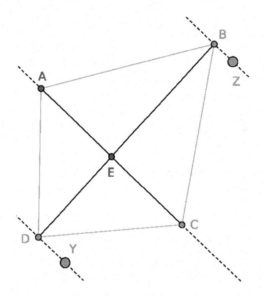

Figure 14.1.31: [1.38, #5]

Extend $\overline{AC}$ to $\overleftrightarrow{AC}$ and construct $\overleftrightarrow{DY}$ and $\overleftrightarrow{BZ}$ such that $\overleftrightarrow{DY} \parallel \overleftrightarrow{AC}$ and $\overleftrightarrow{BZ} \parallel \overleftrightarrow{AC}$; by [1.30], $\overleftrightarrow{DY} \parallel \overleftrightarrow{BZ}$.

Since $\overline{AE} = \overline{EC}$ by hypothesis, by [1.38] $\triangle ABE = \triangle CBE$ and $\triangle ADE = \triangle CDE$. Since $\triangle ADB = \triangle ABE \oplus \triangle ADE$ and $\triangle CDB = \triangle CBE \oplus \triangle CDE$, it follows that $\triangle ADB = \triangle CDB$. Since $ABCD = \triangle ADB \oplus \triangle CDB$, $\overline{BD}$ bisects $ABCD$.

Claim 2: If $ABCD$ is bisected by $\overline{BD}$, then $\overline{BD}$ bisects $\overline{AC}$.

Consider $\triangle ADB$ and $\triangle CDB$: the triangles stand on the same base (BD) but on opposite sides, and $\overline{AC}$ connects their vertices. By [1.37, #1], $\overline{BD}$ bisects $\overline{AC}$.

This completes the proof. □

14.1. SOLUTIONS FOR CHAPTER 1

[1.40] Exercises

1. Triangles with equal bases and altitudes are equal in area.

Proof. Suppose we have two triangles with equal bases and with equal altitudes. Since the altitude of a triangle is the distance between the parallels which contain it, equal altitudes imply that the triangles stand between parallels which are equal distances apart. Therefore [1.37], [1.38], [1.39], and [1.40] prove the claim. □

[1.40] Exercises

2. The segment joining the midpoints of two sides of a triangle is parallel to the base, and the medians from the endpoints of the base to these midpoints each bisect the original triangle. Hence, the two triangles whose base is the third side and whose vertices are the points of bisection are equal in area.

Proof. Construct $\triangle ABC$ with midpoint D on side AB and midpoint E on side AC. Construct $\overline{DE}$, $\overline{DC}$, and $\overline{EB}$. We claim that $\overline{DE} \parallel \overline{BC}$ and

$$\triangle ADC = \triangle BCD = \triangle CBE = \triangle ABE$$

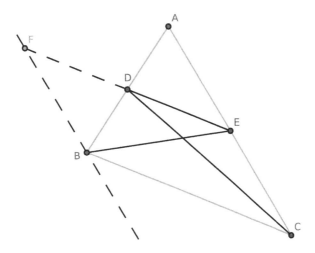

Figure 14.1.32: [1.40, #2]

Construct $\overline{EF} = \overline{DE} \oplus \overline{DF}$ where $\overline{DE} = \overline{DF}$; also construct $\overline{FB}$. Consider $\triangle ADE$ and $\triangle BDF$: by [1.15], $\angle ADE = \angle BDF$; by hypothesis, $\overline{AD} = \overline{BD}$; $\overline{DE} = \overline{DF}$ by construction. By [1.4], $\triangle ADE \cong \triangle BDF$, and so $\angle FBD = \angle DAE$. By [1.29, Cor. 1], $\overline{FB} \parallel \overline{AC}$.

Since $\triangle ADE \cong \triangle BDF$, $\overline{FB} = \overline{AE}$. Because $\overline{AE} = \overline{EC}$ by construction, $\overline{FB} = \overline{EC}$. By the above, $\overline{FB} \parallel \overline{EC}$; by [1.33] $\overline{EF} = \overline{BC}$ and $\overline{EF} \parallel \overline{BC}$. Hence, $\square FECB$ is a parallelogram, and so $\overline{DE} \parallel \overline{BC}$.

By [1.38, #1], $\overline{BE}$ bisects $\triangle ABC$, and so $\triangle ABE = \triangle CBE$. Similarly, $\overline{CD}$ bisects $\triangle ABC$, and so $\triangle ADC = \triangle BDC$.

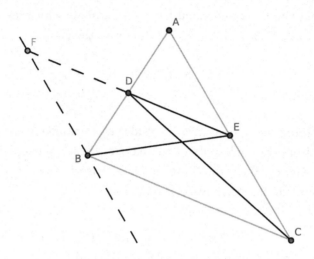

Figure 14.1.33: [1.40, #2]

By applying [1.38, #1] again,

$$\triangle BDC = \frac{1}{2} \cdot \triangle ABC = \triangle CBE$$

By the above, we obtain

$$\triangle ADC = \triangle BCD = \triangle CBE = \triangle ABE$$

which completes the proof. □

14.1. SOLUTIONS FOR CHAPTER 1

[1.40] Exercises

4. The segments which connect the midpoints of the sides of a triangle divide the triangle into four congruent triangles.

Proof. Suppose $\triangle ABC$ has midpoints D on side BC, E on side AC, and F on side AB. Construct segments $\overline{DE}$, $\overline{EF}$, and $\overline{DF}$. We claim that

$$\triangle AEF \cong \triangle ECD \cong \triangle FDB \cong \triangle DFE$$

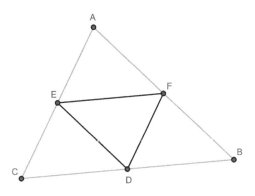

Figure 14.1.34: [1.40, #4]

By [1.40, #2], we find that $\overline{DE} \parallel \overline{AB}$, $\overline{DF} \parallel \overline{AC}$, and $\overline{EF} \parallel \overline{BC}$. By [1.29, Cor. 1], we find that:

$$\angle EDC = \angle EFA = \angle DBF = \angle DEF$$
$$\angle CED = \angle EAF = \angle DFB = \angle EDF$$
$$\angle ECD = \angle AEF = \angle FDB = \angle EFD$$

Since $\square EFCD$ is a parallelogram, $\overline{EF} = \overline{CD}$. Since $\square EFBD$ is a parallelogram, $\overline{EF} = \overline{BD}$, and so

$$\overline{EF} = \overline{CD} = \overline{DB}$$

By multiple applications of [1.26],

$$\triangle AEF \cong \triangle DEF \cong \triangle ECD \cong DBF$$

which proves our claim. □

[1.46] Exercises

1. Two squares have equal side-lengths if and only if the squares are equal in area.

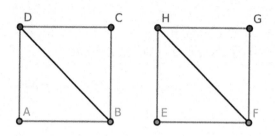

Figure 14.1.35: [1.46, #1]

Proof. Construct squares $\square ABCD$ and $\square EFGH$ as well as $\overline{BD}$ and $\overline{FH}$.

Claim 1: If $\overline{AB} = \overline{EF}$, then $\square ABCD = \square EFGH$.

Consider $\triangle ADB$, $\triangle EHF$: $\overline{AB} = \overline{EF} = \overline{AD} = \overline{EH}$ and $\angle DAB = \angle HEF$ (since each are right angles). By [1.4], $\triangle ABD \cong \triangle EFH$. It follows that $\triangle ABD = \triangle EFH$.

By [1.41], $\square ABCD = \frac{1}{2} \triangle ABD$ and $\square EFGH = \frac{1}{2} \triangle EFH$. By the above, we obtain $\square ABCD = \square EFGH$, which proves our claim.

Claim 2: If $\square ABCD = \square EFGH$, then $\overline{AB} = \overline{EF}$.

By [1.41], $\square ABCD = \frac{1}{2} \triangle ABD$ and $\square EFGH = \frac{1}{2} \triangle EFH$, and so $\triangle ABD = \triangle EFH$. Let b_1 equal the base of $\triangle ABD$, h_1 equal the altitude of $\triangle ABD$, b_2 equal the base of $\triangle EFH$, and h_2 equal the altitude of $\triangle EFH$. Since the area of a triangle $= \frac{1}{2}bh$, we have

$$\frac{1}{2}b_1 h_1 = \frac{1}{2}b_2 h_2$$

Since $\square ABCD$ and $\square EFGH$ are squares, $b_1 = h_1$ and $b_2 = h_2$. Hence, $b_1^2 = b_2^2$, or $\left(\overline{AB}\right)^2 = \left(\overline{EF}\right)^2$. It follows that $\overline{AB} = \overline{EF}$, which completes the proof. $\square$

14.1. SOLUTIONS FOR CHAPTER 1

[1.47] Exercises

4. Find a segment whose square is equal to the sum of the areas of two given squares.

Proof. Let $\square ABCD$ and $\square EFGH$ be the given squares. We wish to construct $\overline{BF}$ such that $(\overline{BF})^2 = \square ABCD + \square EFGH$. Position $\square ABCD$ and $\square EFGH$ such that $C = E$ and $\overline{BC} \perp \overline{DF}$.

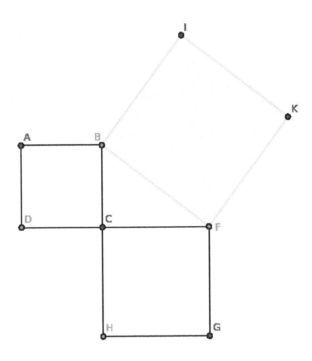

Figure 14.1.36: [1.47, #4]

Construct $\square BFKI$ such that $\square BFKI$ is a square with side-length $\overline{BF}$. By [1.47], $(\overline{BC})^2 + (\overline{CF})^2 = (\overline{BF})^2$. Since $(\overline{BC})^2 = \square ABCD$ and $(\overline{CF})^2 = \square EFGH$, $\overline{BF}$ is the required segment. $\square$

[1.47] Exercises

10. Prove that each of the triangles $\triangle AGK$ and $\triangle BEF$ formed by joining adjacent corners of the squares in [1.47] is equal in area to $\triangle ABC$.

Proof. Construct the polygons as in [1.47] as well as $\overline{KG}$ and $\overline{EF}$. We claim that

$$\triangle KAG = \triangle ABC = \triangle BEF$$

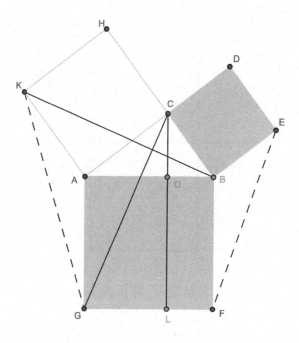

Figure 14.1.37: [1.47, #10]

Recall that the sum of the interior angles of a triangle is π radians in measure. Consider $\triangle ABC$: notice that if $\angle ACB = \frac{\pi}{2}$ radians and $\angle BAC = \gamma$ radians, then $\angle ABC = \frac{\pi}{2} - \gamma$ radians. Since $\angle KAC = \angle GAB = \frac{\pi}{2}$ radians and

$$\angle KAG + \angle KAC + \angle BAC + \angle GAB = 2\pi \text{ radians}$$

it follows that, in radians:

$$\begin{aligned} \angle KAG &= 2\pi - \angle KAC - \angle BAC - \angle GAB \\ &= 2\pi - \frac{\pi}{2} - \gamma - \frac{\pi}{2} \\ &= \pi - \gamma \end{aligned}$$

14.1. SOLUTIONS FOR CHAPTER 1

Similarly

$$\angle FBE = 2\pi - 2 \cdot \frac{\pi}{2} - \left(\frac{\pi}{2} - \gamma\right)$$
$$= \frac{\pi}{2} + \gamma$$

Recall that the general form of the equation of the area of a triangle is:

$$\text{Area} = \frac{1}{2}xy \cdot \sin\theta$$

where θ is the interior angle to sides x and y. So

$$\begin{aligned}
\text{Area } \triangle KAG &= \tfrac{1}{2}AK \cdot AG \cdot \sin(\angle KAG) \\
&= \tfrac{1}{2}AC \cdot AB \cdot \sin(\pi - \gamma) \\
&= \tfrac{1}{2}AC \cdot AB \cdot \sin(\gamma)
\end{aligned}$$

by the properties of the sine function. We also have

$$\begin{aligned}
\text{Area } \triangle ABC &= \tfrac{1}{2}AB \cdot AC \cdot \sin(\angle BAC) \\
&= \tfrac{1}{2}AC \cdot AB \cdot \sin(\gamma) \\
&= \text{Area } \triangle KAG
\end{aligned}$$

Similarly,

$$\begin{aligned}
\text{Area } \triangle BEF &= \tfrac{1}{2}BE \cdot BF \cdot \sin(\angle FBE) \\
&= \tfrac{1}{2}BC \cdot AB \cdot \sin(\tfrac{\pi}{2} + \gamma) \\
&= \tfrac{1}{2}BC \cdot AB \cdot \cos(\gamma)
\end{aligned}$$

and

$$\begin{aligned}
\text{Area } \triangle ABC &= \tfrac{1}{2}AB \cdot BC \cdot \sin(\tfrac{\pi}{2} - \gamma) \\
&= \tfrac{1}{2}BC \cdot AB \cdot \cos(\gamma) \\
&= \text{Area } \triangle BEF
\end{aligned}$$

Therefore $\triangle KAG = \triangle ABC = \triangle BEF$ which proves our claim. $\square$

Chapter 1 exercises

1. Suppose $\triangle_1$ is constructed inside $\triangle_2$ such that each side of $\triangle_2$ passes through one vertex of $\triangle_1$ and each side of $\triangle_2$ is parallel to its opposite side in $\triangle_1$. We claim that $\triangle_2 = 4 \cdot \triangle_1$.

Proof. Construct $\triangle ABC$, $\overleftrightarrow{DF} \parallel \overline{BC}$ such that A is on $\overleftrightarrow{DF}$, $\overleftrightarrow{EF} \parallel \overline{AB}$ such that C is on $\overleftrightarrow{EF}$, and $\overleftrightarrow{DE} \parallel \overline{AC}$ such that B is on $\overleftrightarrow{DE}$. We claim that $\triangle DEF = 4 \cdot \triangle ABC$.

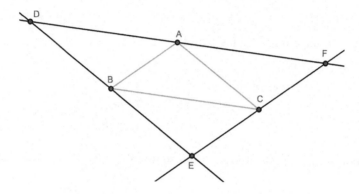

Figure 14.1.38: Chapter 1 exercises, #1

Since $\overleftrightarrow{DF} \parallel \overline{BC}$, by [1.41], $\triangle ABC = \frac{1}{2} \cdot \square AFCB$. Also by [1.41], $\triangle AFC = \frac{1}{2} \cdot \square AFCB$, and so $\triangle ABC = \triangle ACF$. Similarly, $\triangle ABC = \triangle ABD$, or

$$\triangle ABC = \triangle ACF = \triangle ABD$$

Also by [1.41], since $\overleftrightarrow{EF} \parallel \overline{AB}$, $\triangle ABC = \triangle BEC$, or

$$\triangle ABC = \triangle ACF = \triangle ABD = \triangle BEC$$

Since $\triangle DEF = \triangle ABC \oplus \triangle ACF \oplus \triangle ABD \oplus \triangle BEC$, the proof is complete. $\square$

14.1. SOLUTIONS FOR CHAPTER 1

Chapter 1 exercises

8. Construct a triangle given the three medians.

Proof. Suppose we are given the medians of a triangle: $\overline{AE}$, $\overline{BF}$, and $\overline{CD}$. We shall construct $\triangle ABC$.

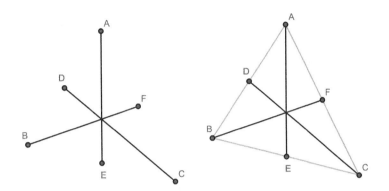

Figure 14.1.39: Chapter 1 exercise #8

Construct $\overline{AB}$, $\overline{AC}$, and $\overline{BC}$. Since $\overline{AE}$ is a median, $\overline{BE} \oplus \overline{EC} = \overline{BC}$ where $\overline{BE} + \overline{EC}$. Similar statements can be made for the remaining sides, *mutatis mutandis*. Hence, $\triangle ABC$ is constructed. □

Chapter 1 exercises

16. Inscribe a rhombus in a triangle having for an angle one angle of the triangle.

Proof. Construct $\triangle ABC$. Let $\angle DAB$ bisect $\angle CAB$ where D is a point on the side BC. Construct $\overrightarrow{AD}$ as well as $\angle ADF$, $\angle ADE$ such that

$$\angle ADF = \angle DAB = \angle ADE$$

We claim that $\square AEDF$ is the required rhombus.

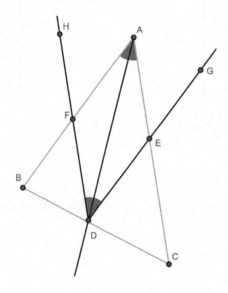

Figure 14.1.40: [Ch. 1 Exercises, #16]

Consider $\triangle DFA$ and $\triangle DEA$: $AF = AE$, $\angle DAF = \angle DAE$ by construction, and the triangles share side AD. By [1.4], $\triangle DFA \cong \triangle DEA$.

Applying [1.6], we obtain

$$DF = FA = AE = ED$$

Since $\angle ADE = \angle DAF$, by [Cor. 1.29.1], $\overline{FA} \parallel \overline{DE}$. Similarly, we can show that $\overline{FD} \parallel \overline{AE}$.

By [Def. 1.29], $\square AEDF$ is a rhombus. Since $\square AEDF$ is inscribed in $\triangle ABC$ and $\triangle ABC$ shares $\angle BAC$ with $\square AEDF$, our proof is complete. $\square$

14.2 Solutions for Chapter 2

[2.4] Exercises

2. If from the right angle of a right triangle a perpendicular falls on the hypotenuse, its square equals the area of the rectangle contained by the segments of the hypotenuse.

Proof. Construct right triangle $\triangle ABC$ where $\angle BAC$ is a right angle. Construct segment $\overline{AD}$ such that $\overline{AD} \perp \overline{BC}$. We claim that $(\overline{AD})^2 = \overline{DB} \cdot \overline{DC}$.

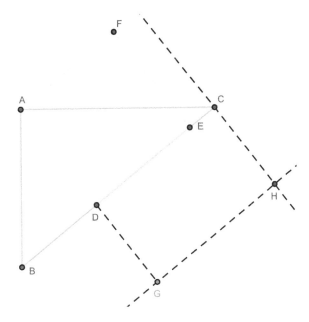

Figure 14.2.1: [2.4, #2]

Construct rectangle $\square DCHG$ where $\overline{BD} = \overline{DG}$. (Note: geometrically, we claim that $(\overline{AD})^2 = \square DCHG$.)

By [1.47], we find that
$$(\overline{AD})^2 + (\overline{DC})^2 = (\overline{AC})^2$$
$$(\overline{AD})^2 + (\overline{DB})^2 = (\overline{AB})^2$$

as well as
$$(\overline{AB})^2 + (\overline{AC})^2 = (\overline{DB} + \overline{DC})^2$$
$$= (\overline{DB})^2 + 2 \cdot \overline{DB} \cdot \overline{DC} + (\overline{DC})^2$$

Hence,
$$(\overline{AD})^2 + (\overline{DC})^2 + (\overline{AD})^2 + (\overline{DB})^2 = (\overline{AB})^2 + (\overline{AC})^2$$
$$2 \cdot (\overline{AD})^2 + (\overline{DC})^2 + (\overline{DB})^2 = (\overline{DB})^2 + 2 \cdot \overline{DB} \cdot \overline{DC} + (\overline{DC})^2$$
$$2 \cdot (\overline{AD})^2 = 2 \cdot \overline{DB} \cdot \overline{DC}$$
$$(\overline{AD})^2 = \overline{DB} \cdot \overline{DC}$$

which completes the proof. □

[2.4] Exercises

9. Prove [Cor. 2.4.4]: The square on a segment is equal in area to four times the square on its half.

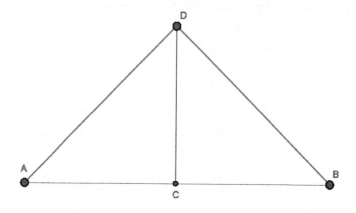

Figure 14.2.2: [2.4, Cor. 3]

Proof. Suppose we have $\triangle ABD$ such that $\overline{AB} = 2 \cdot \overline{AC}$, $\overline{CD} \perp \overline{AB}$, and $\overline{CD} = \overline{AC}$. We claim that $(\overline{AB})^2 = 4 \cdot (\overline{AC})^2$.

Let $\overline{AC} = x$. Then $\overline{AB} = 2x$. It follows that

$$(\overline{AB})^2 = (2x)^2 = 4x^2 = 4 \cdot (\overline{AC})^2$$

This proves our claim. □

[2.6] Exercises

7. Give a common statement which will include [2.5] and [2.6].

Proof. Construct $\overleftrightarrow{AB}$ and on $\overline{AB}$, locate midpoint C. Choose a point D on $\overleftrightarrow{AB}$ such that D is neither A, B, nor C. We have two cases:

1) D is between A and B. By [2.5], $\overline{AD} \cdot \overline{DB} + (\overline{CD})^2 = (\overline{CB})^2$

2) D is not between A and B. By [2.6], $\overline{AD} \cdot \overline{DB} + (\overline{CB})^2 = (\overline{CD})^2$

This completes the proof. □

14.2. SOLUTIONS FOR CHAPTER 2

[2.11] Exercises

3. If $\overline{AB}$ is cut in "extreme and mean ratio" at H, prove that

(a) $(\overline{AB})^2 + (\overline{BH})^2 = 3 \cdot (\overline{AH})^2$

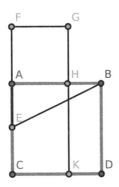

Figure 14.2.3: [2.11]

Proof. Using the construction from [2.11], $x = -\frac{a}{2}(1 \pm \sqrt{5})$. (We may ignore results where $x \leq 0$.) Since $\overline{AB} = a$, $\overline{BH} = a - x$, and $\overline{AH} = x$, notice that $x^2 = \frac{a^2}{2}\left(3 \pm \sqrt{5}\right)$ and $(a-x)^2 = \frac{a^2}{2}\left(7 \pm 3\sqrt{5}\right)$. Or,

$$\begin{aligned}
(\overline{AB})^2 + (\overline{BH})^2 &= a^2 + (a-x)^2 \\
&= a^2 + \frac{a^2}{2}\left(7 \pm 3\sqrt{5}\right) \\
&= a^2\left(\frac{9}{2} \pm \frac{3}{2}\sqrt{5}\right) \\
&= \frac{3a^2}{2}\left(3 \pm \sqrt{5}\right) \\
&= 3x^2 \\
&= 3 \cdot (\overline{AH})^2
\end{aligned}$$

which completes the proof. □

Chapter 2 exercises

15. Any rectangle is equal in area to half the rectangle contained by the diagonals of squares constructed on its adjacent sides.

Proof. Construct rectangle $\square ADCB$, squares $\square GABH$ and $\square BCFE$, and diagonals $\overline{GB}$ and $\overline{BF}$. We claim that $\frac{1}{2} \cdot \overline{GB} \cdot \overline{BF} = \overline{AB} \cdot \overline{BC}$.

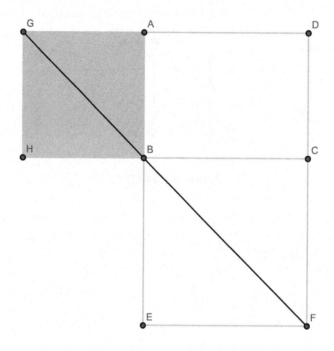

Figure 14.2.4: [Ch. 2 Exercises, #15]

Let $\overline{AB} = x$ and $\overline{BC} = y$. By [1.47], it follows that $\overline{GB} = x\sqrt{2}$ and $\overline{BF} = y\sqrt{2}$. Then

$$\begin{aligned}
\tfrac{1}{2} \cdot \overline{GB} \cdot \overline{BF} &= \tfrac{1}{2} \cdot 2xy \\
&= xy \\
&= \overline{AB} \cdot \overline{BC}
\end{aligned}$$

which completes the proof. □

14.3 Solutions for Chapter 3

[3.3] Exercises

5. Prove [3.3, Cor. 4]: The line joining the centers of two intersecting circles bisects their common chord perpendicularly.

Proof. Construct the figures from [1.1, #2]. We claim that $\overline{AB} \perp \overline{CF}$ and that $\overline{AB}$ bisects $\overline{CF}$.

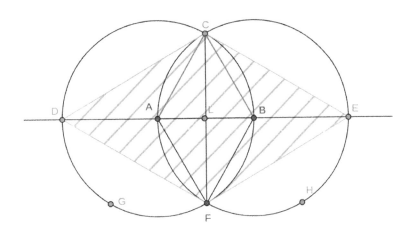

Figure 14.3.1: [1.1, #2] and [3.3, #5]

From the proof of [1.1, #2], $\angle ALC = \angle ALF$. Since the angles are adjacent, they are right angles; otherwise, $\overline{CF}$ would not be a side of $\triangle CDF$ and $\triangle CEF$. Hence, $\overline{AB} \perp \overline{CF}$.

Notice that $\triangle ACL$ and $\triangle AFL$ are right triangles. By [1.47], $\left(\overline{AL}\right)^2 + \left(\overline{LC}\right)^2 = \left(\overline{AC}\right)^2$ and $\left(\overline{AL}\right)^2 + \left(\overline{LF}\right)^2 = \left(\overline{AF}\right)^2$.

Since $\overline{AC}$ and $\overline{AF}$ are radii of $\odot A$, $\overline{AC} = \overline{AF}$. Hence, $\left(\overline{AL}\right)^2 + \left(\overline{LC}\right)^2 = \left(\overline{AL}\right)^2 + \left(\overline{LF}\right)^2$, which simplifies to $\overline{LC} = \overline{LF}$. Since $\overline{CF} = \overline{LC} \oplus \overline{LF}$, $\overline{AB}$ bisects $\overline{CF}$, which completes the proof. $\square$

[3.5] Exercises

2. Two circles cannot have three points in common without coinciding.

Proof. Suppose instead that two circles ($\odot C$, $\odot A$) have three points in common (E, F, and G) and do not coincide.

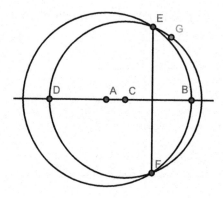

Figure 14.3.2: [3.5, #2]

By [3.3, Cor. 4], the line joining the centers of two intersecting circles ($\overleftrightarrow{AC}$) bisects their common chord perpendicularly; hence, $\overleftrightarrow{AC} \perp \overline{EF}$.

Similarly, $\overleftrightarrow{AC}$ bisects $\overline{EG}$. But $\overline{EG}$ can be constructed so that $\overline{EG}$ and $\overleftrightarrow{AC}$ do not intersect, a contradiction. Therefore, $\odot C$ and $\odot A$ coincide. □

[3.13] Exercises

3. Suppose two circles touch externally. If through the point of intersection any secant is constructed cutting the circles again at two points, the radii constructed to these points are parallel.

Proof. Suppose $\odot A$ and $\odot C$ touch at point B. By [3.13], these circles touch only at B. Construct secant $\overline{DBE}$. We claim that $\overline{AD} \parallel \overline{CE}$.

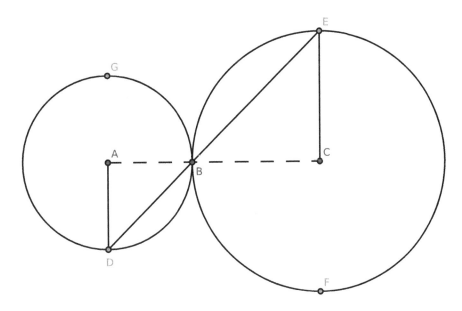

Figure 14.3.3: [3.13, #3]

Construct $\overline{AC}$. By [3.12], $\overline{AC}$ intersects B.

Consider $\triangle ABD$ and $\triangle CBE$: $\angle ABD = \angle CBE$ by [1.15]; since each triangle is isosceles, $\angle ADB = \angle CEB$. By [1.29, Cor. 1], $\overline{AD} \parallel \overline{CE}$. This proves our claim. □

Corollary. *If two circles touch externally and through the point of intersection any secant is constructed cutting the circles again at two points, the diameters constructed to these points are parallel.*

[3.13] Exercises

4. Suppose two circles touch externally. If two diameters in these circles are parallel, the line from the point of intersection to the endpoint of one diameter passes through the endpoint of the other.

Proof. Suppose $\odot A$ and $\odot C$ touch at point B. By [3.13], these circles touch only at B. Construct $\overline{AC}$, and construct diameters $\overline{DE}$ and $\overline{FG}$ such that $\overline{DE} \parallel \overline{FG}$ and $\overline{DE} \perp \overline{AC}$. It follows that $\overline{AC} \perp \overline{FG}$. We claim that $\overleftrightarrow{DB}$ intersects G.

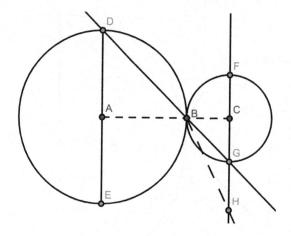

Figure 14.3.4: [3.13, #4]

Suppose that $\overleftrightarrow{DB}$ does not intersect G. Extend $\overline{FG}$ to $\overleftrightarrow{FG}$ and suppose that $\overleftrightarrow{DB}$ intersects $\overleftrightarrow{FG}$ at H.

Consider $\triangle ABD$ and $\triangle CBH$: by hypothesis, $\angle DAB = \angle HCB$; by [1.15], $\angle ABD = \angle CBH$; applying [1.32], we obtain $\angle ADB = \angle CHB$. That is, $\triangle ABD$ and $\triangle CBH$ are equiangular.

However, $\triangle ABD$ is an isosceles triangle and $\triangle CBH$ is not since $\overline{BC} = \overline{CG}$ and $\overline{CG} < \overline{CH}$; hence, the triangles are not equiangular, a contradiction.

An equivalent contradiction is obtained if $\overleftrightarrow{DB}$ intersects $\overleftrightarrow{FG}$ at any point other than G, *mutatis mutandis*. This proves our claim. □

[3.16] Exercises

1. If two circles are concentric, all chords of the larger circle which touch the smaller circle are equal in length.

Proof. Construct $\bigcirc A_1$ with radius $\overline{AB}$ and $\bigcirc A_2$ with radius $\overline{AC}$. On $\bigcirc A_2$, construct chord $\overline{DE}$ such that $\overline{DE}$ touches $\bigcirc A_1$ at B. Also on $\bigcirc A_2$, construct chord $\overline{HJ}$ such that $\overline{HJ}$ touches $\bigcirc A_1$ at G. We claim that $\overline{DE} = \overline{HJ}$.

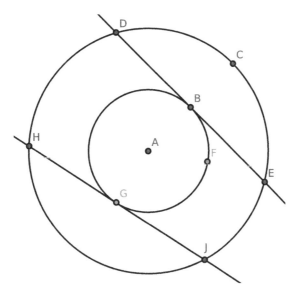

Figure 14.3.5: [3.16, #1]

Notice that $\overline{AG} = \overline{AB}$ since each are radii of $\bigcirc A_1$. By [3.16], $\overline{DE}$ and $\overline{HJ}$ have no other points of intersection with $\bigcirc A_1$. Hence, $\overline{DE}$ and $\overline{HJ}$ are equal distance from A, the center of $\bigcirc A_2$. By [3.14], $\overline{DE} = \overline{HJ}$. This proves our claim. □

CHAPTER 14. SOLUTIONS

[3.30] Exercises

1. Suppose that $ABCD$ is a semicircle with diameter $\overline{AD}$ and a chord $\overline{BC}$. Extend $\overline{BC}$ to $\overrightarrow{BC}$ and $\overline{AD}$ to $\overrightarrow{AD}$, and suppose each ray intersects at E. Prove that if $\overline{CE}$ is equal in length to the radius of $ABCD$, then arc $\overline{AB} = 3 \cdot \overline{CD}$.

Proof. Our hypothesis and claim are stated above.

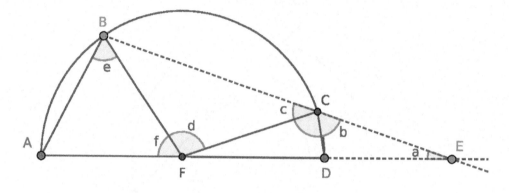

Figure 14.3.6: [3.30, #1]

Construct $\overline{CD}, \overline{CF}, \overline{FB}$, and $\overline{AB}$. Notice that

$$\overline{CE} = \overline{DF} = \overline{CF} = \overline{BF} = \overline{AF}$$

By [1.5], we find that $\angle CEF = \angle CFE$, $\angle FCB = \angle FBC$, and $\angle FBA = \angle FAB$.

Define $\angle CEF = a$, $\angle ECF = b$, $\angle FCB = c$, $\angle CFB = d$, $\angle FBA = e$, and $\angle BFA = f$. Using linear algebra, we obtain:

$$\begin{aligned} b + c &= 180 \\ 2a + b &= 180 \\ 2c + d &= 180 \\ 2e + f &= 180 \\ a + d + f &= 180 \\ a + c + 2e &= 180 \end{aligned}$$

where the RHS is in degrees. In matrix form:

$$\begin{bmatrix} 0 & 1 & 1 & 0 & 0 & 0 & | & 180 \\ 2 & 1 & 0 & 0 & 0 & 0 & | & 180 \\ 0 & 0 & 2 & 1 & 0 & 0 & | & 180 \\ 0 & 0 & 0 & 0 & 2 & 1 & | & 180 \\ 1 & 0 & 0 & 1 & 0 & 1 & | & 180 \\ 1 & 0 & 1 & 0 & 2 & 0 & | & 180 \end{bmatrix}$$

14.3. SOLUTIONS FOR CHAPTER 3

In reduced row echelon form, this becomes:

$$\begin{bmatrix} 1 & 0 & 0 & 0 & 0 & 0 & | & 20 \\ 0 & 1 & 0 & 0 & 0 & 0 & | & 140 \\ 0 & 0 & 1 & 0 & 0 & 0 & | & 40 \\ 0 & 0 & 0 & 1 & 0 & 0 & | & 100 \\ 0 & 0 & 0 & 0 & 1 & 0 & | & 60 \\ 0 & 0 & 0 & 0 & 0 & 1 & | & 60 \end{bmatrix}$$

Or, $\angle CEF = 20°$, $\angle ECF = 140°$, $\angle FCB = 40°$, $\angle CFB = 100°$, $\angle FBA = 60°$, and $\angle BFA = 60°$.

Since $\triangle CEF$ is isosceles, $\angle CFD = \angle CFE = 20°$, and so $\angle BFA = 3 \cdot \angle CFD = 60°$. By [7.29, Cor. 1], it follows that $\overline{AB} = 3 \cdot \overline{CD}$. □

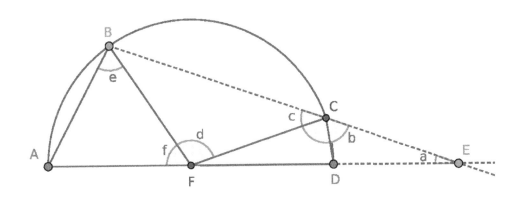

14.4 Solutions for Chapter 4

[4.4] Exercises

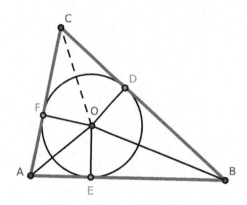

Figure 14.4.1: [4.4, #1]

1. In [4.4]: if $\overline{OC}$ is constructed, prove that the angle $\angle ACB$ is bisected. Hence, we prove the existence of the *incenter* of a triangle.

Proof. Consider $\triangle OFC$ and $\triangle CDO$: by the proof of [4.4], $\angle OFC = \angle CDO$ since each are right angles. By [1.47],

$$\left(\overline{OF}\right)^2 + \left(\overline{FC}\right)^2 = \left(\overline{OC}\right)^2$$
$$\left(\overline{OD}\right)^2 + \left(\overline{DC}\right)^2 = \left(\overline{OC}\right)^2$$

Since $\overline{OF} = \overline{OD}$ (each are radii of $\odot O$), $\overline{FC} = \overline{DC}$. By [1.8], $\triangle OFC \cong \triangle CDO$, and so $\angle OCF = \angle OCD$.

It follows that $\overline{OC}$ bisects $\angle ACB$, and therefore O is the incenter of $\triangle ABC$; that is, O is the point of intersection of the bisectors of the three internal angles of $\triangle ABC$. □

14.4. SOLUTIONS FOR CHAPTER 4

[4.5] Exercises

1. Prove that the three altitudes of a triangle ($\triangle ABC$) are concurrent. (This proves the existence of the *orthocenter* of a triangle.)

Proof. Construct $\triangle ABC$; also construct altitudes $\overline{AE}$, $\overline{BD}$, and $\overline{CF}$. We wish to show these altitudes are concurrent.

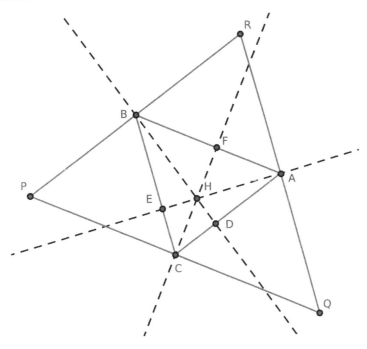

Figure 14.4.2: [4.5, #1]

Through vertex A, construct $\overleftrightarrow{QR}$ such that $\overleftrightarrow{QR} \parallel \overline{BC}$. Similarly, construct $\overleftrightarrow{PQ} \parallel \overline{AB}$ through C and $\overleftrightarrow{PR} \parallel \overline{AC}$ through B. Notice that the segments $\overline{QR}$, $\overline{PQ}$, and $\overline{PR}$ constitute $\triangle PQR$.

We have also constructed $\square RACB$, $\square QABC$, and $\square PBAC$. It follows that $\overline{AR} = \overline{BC}$ and $\overline{AQ} = \overline{BC}$, or $\overline{AR} = \overline{AQ}$. Hence, A is the midpoint $\overline{QR}$; similarly, B is the midpoint $\overline{PR}$ and C is the midpoint of $\overline{PQ}$.

Since $\overline{AE} \perp \overline{BC}$ and $\overline{BC} \parallel \overline{QR}$, $\overline{AE} \perp \overline{QR}$; or, $\overline{AE}$ is the perpendicular bisector of $\overline{QR}$. Similarly, $\overline{BD}$ is the perpendicular bisector of $\overline{RP}$, and $\overline{CF}$ is the perpendicular bisector of $\overline{PQ}$. All are concurrent at the circumcenter of $\triangle PRQ$ by [4.5].

Since these segments are also the altitudes of $\triangle ABC$, the proof is complete. $\square$

[4.7] Exercises

1. Prove [Cor. 4.7.1]: the circumscribed square, □EHGF, has double the area of the inscribed square, □BCDA.

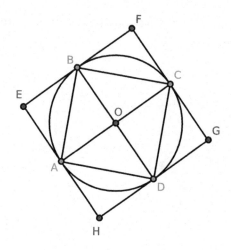

Figure 14.4.3: [4.7]

Proof. Consider □BFCO: by [1.34], $\overline{BC}$ bisects □BFCO, and so □BFCO = $2 \cdot \triangle OBC$.

Consider $\triangle OBA$ and $\triangle ODC$: $\angle AOB = \angle DOC$ by [1.15] and the adjacent sides to these angles are equal since they are radii of $\odot O$. By [1.4], $\triangle OBA \cong \triangle ODC$.

Consider $\triangle OBA$ and $\triangle OBC$: $\angle AOB = \angle COB$ since $\overline{AC}$ is a segment and the adjacent sides to these angles are equal since each are radii of $\odot O$. Again by [1.4], $\triangle OBA \cong \triangle OBC$. It follows that $\triangle OBA \cong \triangle OBC \cong \triangle ODC \cong \triangle OAD$, and so each of these triangles is equal in area.

Consider $\triangle EBA$ and $\triangle OBA$: $\overline{EB} = \overline{OA} = \overline{AE} = \overline{OB}$ since □EBOA is a square, and each triangle shares $\overline{AB}$. By [1.8], $\triangle EBA \cong \triangle OBA$, and so $\triangle EBA = \triangle OBA$.

Since □EBOA = $\triangle EBA \oplus \triangle OBA$, □EBOA = $2 \cdot \triangle EBA$.

By the proof of [4.7], □EBOA = □BFCO = □AODH = □OCGD. Since

$$\square EHGF = \square EBOA \oplus \square BFCO \oplus \square AODH \oplus \square OCGD$$

it follows that □EHGF = $4 \cdot$ □EBOA = $8 \cdot \triangle EBA$.

Notice that □BCDA = $4 \cdot \triangle EBA$, and so □EHGF = $2 \cdot$ □BCDA, which completes the proof. □

14.4. SOLUTIONS FOR CHAPTER 4

[4.10] Exercises

1. Prove that $\triangle ACD$ is an isosceles triangle whose vertical angle is equal to three times each of the base angles.

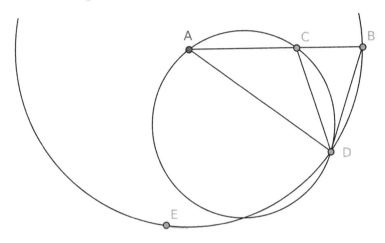

Figure 14.4.4: [4.10]

Proof. We claim that $\triangle ACD$ is isosceles where

$$\angle ACD = 3 \cdot \angle DAC = 3 \cdot \angle ADC$$

Since $\angle BDA = \angle DBA = 2 \cdot \angle DAB$ by the proof of [4.10], we must have that

$$\angle DAB = 36°$$
$$\angle DBA = 72°$$
$$\angle BDA = 72°$$

Notice that

$$\angle DAB = \angle DAC = \angle ADC = 36°$$

By [1.13], $\angle ACD = 108°$, and since $36 \cdot 3 = 108$, the proof is complete. □

14.5 Solutions for Chapter 5

Chapter 5 Exercises

2. If four numbers are proportionals, their squares, cubes, etc., are proportionals.

Proof. Let a, b, x, and y be natural numbers such that

$$\frac{a}{b} = \frac{x}{y}$$

We claim that

$$\left(\frac{a}{b}\right)^n = \left(\frac{x}{y}\right)^n$$

where $n \geq 1$ is a positive integer.

The equality holds for $n = 1$ by assumption. Assume that the equality holds for $n = k$:

$$\left(\frac{a}{b}\right)^k = \left(\frac{x}{y}\right)^k$$

Then

$$\left(\frac{a}{b}\right) \cdot \left(\frac{a}{b}\right)^k = \left(\frac{a}{b}\right) \cdot \left(\frac{x}{y}\right)^k$$
$$\left(\frac{a}{b}\right) \cdot \left(\frac{a}{b}\right)^k = \left(\frac{x}{y}\right) \cdot \left(\frac{x}{y}\right)^k$$
$$\left(\frac{a}{b}\right)^{k+1} = \left(\frac{x}{y}\right)^{k+1}$$

And so the equality holds for $n = k+1$, and therefore it holds for $n \geq 1$. The proves our claim. □

14.6 Solutions for Chapter 6

[6.7] Exercises

3. Prove the Transitivity of Similar Triangles, i.e., if $\triangle ABC \sim \triangle DEF$ and $\triangle DEF \sim \triangle GHI$, then $\triangle ABC \sim \triangle GHI$.

Proof. Suppose $\triangle ABC \sim \triangle DEF$ and $\triangle DEF \sim \triangle GHI$. We wish to show that $\triangle ABC \sim \triangle GHI$.

By [Cor. 6.4.1], $\triangle ABC$ and $\triangle DEF$ are equiangular; also, $\triangle DEF$ and $\triangle GHI$ are equiangular. Clearly, $\triangle ABC$ and $\triangle GHI$ are equiangular.

Applying [Cor. 6.4.1] again, $\triangle ABC \sim \triangle GHI$. This completes the proof. □

Questions? Comments? Did you find an error?

Email me at: dpcallahan@protonmail.com

Make sure to include the version number, which can be found at the beginning of this document.

Made in the USA
Las Vegas, NV
15 February 2023

67594617R00208